SpringerBriefs in Materials

For further volumes:
http://www.springer.com/series/10111

Edson Roberto Leite · Caue Ribeiro

Crystallization and Growth of Colloidal Nanocrystals

 Springer

Edson Roberto Leite
Centro de Ciências Exatas e de
Tecnologia
Universidade Federal de São Carlos
São Carlos São Paulo, Brazil
edson.leite@pq.cnpq.br

Caue Ribeiro
Agropecuária (EMBRAPA)
Empresa Brasileira de Pesquisa
São Carlos São Paulo, Brazil
caue@cnpdia.embrapa.br

ISSN 2192-1091   e-ISSN 2192-1105
ISBN 978-1-4614-1307-3   e-ISBN 978-1-4614-1308-0
DOI 10.1007/978-1-4614-1308-0
Springer New York Dordrecht Heidelberg London

Library of Congress Control Number: 2011940975

Printed on acid-free paper

Springer is part of Springer Science+Business Media (www.springer.com)

Preface

In the last two decades, the study and development of nanomaterials has made impressive progress, particularly in the chemical synthesis of inorganic nanocrystals via the colloidal process. The synthesis of a nanocrystal is not a simple task, because it requires control over the chemical composition and the purity, size, size distribution, and shape of the crystallographic phase, as well as the chemical functionality of the nanocrystal surface. Today, the development of synthetic routes to produce nanocrystals with controlled size, size distribution, and shape is directly correlated with the ability to control the nucleation and growth process. Detailed investigations of the nucleation and growth process of nanocrystals in the colloidal state, and an understanding of the parameters that affect crystal growth and shape development, are fundamental to the tailoring of new types of nanostructured functional materials and even for the development of chemical protocols that allow for the processing of nanocrystals with high yield and reproducibility. The main challenge and goal of this book is to present readers with the fundamental principles of nucleation and growth process of nanocrystals.

We have attempted to accomplish this goal in a dynamic yet not exhaustive way. Nonclassical mechanisms are explored, and information is developed in a progressive fashion. For instance, the fundamental concepts in Chap. 2 are applied in Chaps. 3, 4, and 5, and so on. In order to avoid a long and exhaustive book, the contents of the various chapters are based on our personal experience in this research field as acquired in the last 10 years.

The book starts with a general and introductory chapter, where we show why it is important to understand the nucleation and growth process of nanocrystals. After this short chapter, we discuss the basic principles of thermodynamics and colloidal science in Chap. 2. Chapter 3 is dedicated to the classical nucleation and growth process, and focuses on the Ostwald ripening mechanism. In Chap. 4, the nonclassical crystallization process is presented. This subject has been experiencing increased attention from chemists and physicists in recent years, and for materials chemists in particular, mechanism such as oriented attachment offers new opportunities for materials design. A correlation between oriented attachment and colloidal state is also presented in this chapter. Chapter 5 describes special cases of nonclassical growth process, such as oriented attachment in the presence of a solid–solid interface. This chapter presents new ideas about this mechanism and shows that it is important even in a hampered system. The last chapter of the book summarizes the current knowledge and the future challenges in the field of nucleation and growth of colloidal nanocrystals.

We are thankful to our current and past group members for their great scientific work. The interaction with graduate students and postdoctoral researchers has shown us the importance of a book that lays out, clearly and succinctly, the basic principles of the nucleation and growth process.

São Carlos, São Paulo   Edson Roberto Leite
São Carlos, São Paulo   Caue Ribeiro

Contents

1 Introduction.. 1
    References.. 4

2 Basic Principles: Thermodynamics and Colloidal Chemistry 7
    2.1   Thermodynamics Basic Principles.............................. 7
    2.2   Colloidal Chemistry Basic Principles.......................... 12
    References.. 17

3 Classical Crystallization Model: Nucleation and Growth........... 19
    3.1   The Bubble Model 20
    3.2   Homogeneous Nucleation... 22
    3.3   The Nucleation Rate 28
    3.4   Growth in Reactional Steps...................................... 30
    3.5   Heterogeneous Nucleation 33
    3.6   Deviations of the Models: The Determination of γ 34
    3.7   Crystal Growth... 36
    3.8   Ostwald Ripening 36
    References.. 39

4 Oriented Attachment and Mesocrystals 45
    4.1   A Qualitative Analysis of the Oriented Attachment (OA)
          Mechanism 45
          4.1.1   OA in a Dispersed Colloidal State........................ 46
          4.1.2   OA in a Weakly Flocculated Colloidal State.............. 48
    4.2   Quantitative Description of the OA Mechanism 53
          4.2.1   Oriented Attachment in the Dispersed Colloidal
                  State: A Quantitative Description.......................... 53
    References.. 65

5 Oriented Attachment (OA) with Solid–Solid Interface 69
  5.1 Quantitative Analysis of the GRIGC Mechanism 69
  5.2 The Self-recrystallization Process 74
  5.3 Crystal Growth and Phase Transformation 76
  References 78

6 Trends and Perspectives in Nanoparticles Synthesis 83
  6.1 Trends in the Synthesis of Transition Metal Oxides 83
  6.2 Trends in the Application of Metal Oxides
      with Controlled Shapes and Reactive Surfaces 89
  References 90

Index 93

Chapter 1
Introduction

The transmission electron microscopy image in Fig. 1.1a shows well-dispersed magnetite particles with a dimension below 10 nm, i.e., nanoparticles. The black arrows in this figure indicate the presence of equiaxial (almost spherical) particles and well-faceted (triangular-shaped) nanoparticles. A high-resolution transmission electron microscopy analysis (see Fig. 1.1b, c) reveals that each particle is formed by a single crystalline domain; i.e., each particle is a single crystal and both particles have the same crystalline structure despite a different shape. Several questions arise from an analysis of these figures: Why do nanoparticles have different shapes and good dispersion? Is it possible to control the shape and size of nanoparticles? The driving force that motivates this book is basically to answer or at least supply fundamental knowledge to answer these questions.

A nanocrystal is a crystalline entity with a specific shape and specific number of atoms. A nanocrystal is precisely defined as a nanoparticle with only one crystalline domain (see Fig. 1.1). The unusual chemical and physical properties of nanocrystals are determined not only by the large number of atoms on the surface but also by the crystallographic structure of the particle. For a spherical nanocrystal of radius r, the ratio between the number of atoms in the surface (N_S) to the number of atoms in the volume or bulk of the nanocrystal (N_V) is

$$N_s/N_v = 3\rho_s/r\rho_v, \tag{1.1}$$

where ρ_S is the density of atoms on the surface, and ρ_V is the bulk density of the crystal. Applying Eq. 1.1 for gold nanocrystals and considering that the particle surface is composed of (100) and (111) planes, over 30% of gold atoms are on the surface when the crystal size is smaller than 3 nm.

Nanocrystals generally display properties that differ from bulk material properties. The correlation between properties and particle size has been known since the nineteenth century when M. Faraday demonstrated that the color of colloidal Au particles can be modified and proposed that the effect was related to the particle size [1]; however, an effective control over synthesis and properties was

E.R. Leite and C. Ribeiro, *Crystallization and Growth of Colloidal Nanocrystals*,
SpringerBriefs in Materials, DOI 10.1007/978-1-4614-1308-0_1,
© Edson Roberto Leite and Caue Ribeiro 2012

Fig. 1.1 (**a**) Transmission electron microscopy image of magnetite nanoparticles with different morphologies. (**b**) High-resolution transmission electron microscopy image of an equiaxial nanoparticle. (**c**) High-resolution transmission electron microscopy image of a triangular nanoparticle

achieved only in the last two decades of the twentieth century. The property of a material is usually substantially modified in 1–10 nm sizes. These changes (quantum size effects) are directly related to the type of chemical bond in the crystal [2]. The literature provides several examples of properties (e.g., magnetic and optical), melting point, specific heat, and surface reactivity which can be affected by particle size [3–9].

A key factor in the technological application of these properties is the ability to control the size and shape of crystals and the arrangement of particles [10, 11]. The range of possible new technologies is immense, including energy conversion [12–17], catalysis and sensors [18, 19], ultrahigh-density data storage media [20–22], nanoparticle light-emitting diodes [23, 24], and special pigments [25]. However, the future of these new technologies is strictly dependent on the development of synthetic routes to process metal, metal oxides, and semiconductor nanoparticles, as well as processes that allow such nanoparticles to be manipulated and controlled. In this technology, a deeper comprehension of phenomena such as nucleation and growth will be fundamental to define correct strategies to tailor the desired sizes or shapes of any system. The importance of the concepts appears in several other areas such as the production of single crystals from melts [26], the regular precipitation of single crystals in a reaction medium [27], control of glass forming and crystallization in glasses [28–30], structural control in recrystallization [7],

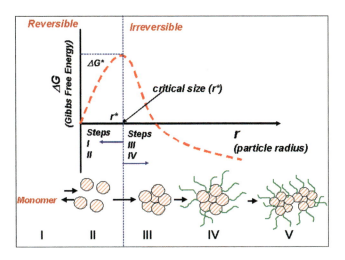

Fig. 1.2 Schematic diagram of the nucleation and growth process showing the five steps and the dependence of the Gibbs free energy (G) with the crystal size. Below the critical radius (r^*), a reversible process occurs; for $r > r^*$, the process becomes irreversible

and solid sintering [31–35], etc. This knowledge was stabilized in the last century, but the implications in nanocrystal synthesis are still basic research in several cases.

In the last two decades, the study and development of nanomaterials has made impressive progress, particularly in the chemical synthesis of inorganic nanocrystals via the colloidal process [36, 37]. An important point to be considered is the fact that during the genesis of nanocrystals by a colloidal approach, a phase-separation process takes place which is generally associated with a nucleation and growth process [38–40]. For didactic purposes, the nucleation and growth process can be divided into five steps (see Fig. 1.2). The first step consists of the reaction of suitable precursors (monomers) which results in a desirable compound (e.g., metal atoms, a semiconductor or an oxide compound). This compound will interact (step II), resulting in a cluster or monomer growth process. The growth mechanism that promotes the growth of the cluster is not yet established, and steps I and II are reversible. When the cluster grows to a critical size (step III), the process becomes irreversible (a thermodynamic condition), and the crystal size can be controlled with the aid of stabilizers (step IV). In a first approximation, the shape of the crystal is controlled by energetic arguments (a thermodynamic condition). In fact, the cluster will grow in a geometric arrangement to minimize the surface energy. In general, the Wulff construction is an interesting way to predict the shape of the nanocrystal [41, 42]. In this step, one can interfere in the process by promoting, for example, the preferential adsorption/desorption of a stabilizer which results in an anisotropic crystal [36, 37, 43]. In fact, the size and shape of the nanocrystal can be controlled using a suitable combination of stabilizers and solvents (a kinetic condition). The last step (step V) consists of the interaction of the nanocrystals formed in step IV in order to produce larger structures. This nanocrystal-based self-assembly process is governed by particle-particle and particle-solvent interactions. In this step, the formation of agglomerated (disordered assembly of nanocrystals) or even the formation of mesocrystals may occur [44]. Needless to say, the above-described process is ideal. One of the major obstacles to achieving good control of the colloidal synthesis of nanocrystals is to separate the nucleation event from the growth process.

The literature shows different methods to produce nanocrystals based on the separation of the nucleation and growth processes, which facilitates better control of the size and morphology of the materials produced. Thus, this book will discuss the fundamentals of nucleation and growth theories and how those theories can be applied in nanocrystal synthesis. An important part of the present work is to introduce new concepts regarding the nonclassical crystallization process as well as new concepts of crystal growth based on the attachment of nanocrystals or clusters (oriented attachment). Very recent research [45, 46] suggests that the oriented attachment mechanism is a fundamental process that can explain the initial growth process after the nucleation process occurs in a condition where the clusters and monomer concentration are significant. Thus, a new term starts to appear in the literature: the monomer attachment.

References

1. Faraday, M.: The Bakerian lecture: Experimental relations of gold (and other metals) to light. Philos. Trans. R. Soc. **147**, 145 (1857)
2. Mulvaney, P.: Not all that's gold does glitter. MRS Bull. **26**, 1009 (2001)
3. Interrante, L.V., Hampden-Smith, M.J.: Chemistry of advanced materials. Willey-VCH, New York (1998)
4. Alivisatos, A.P.: Perspectives on the physical chemistry of semiconductor nanocrystals. J. Phys. Chem. **100**, 13226 (1996)
5. Morup, S.: Studies of superparamagnetism in samples of ultrafine particles. In: Nanomagnetism, pp. 93–99. Kluwer Academic Publishers, Boston (1993)
6. O'Grady, K., Chantrell, R.W.: Magnetic properties: Theory and experiments. In: Magnetic Properties of Fine Particles. Elsevier, Amsterdam (1992)
7. Gleiter, H.: Nanostructured materials: Basic concepts and microstructure. Acta Mater. **48**, 1 (2000)
8. Gratzel, M.: Photoelectrochemical cells. Nature **414**, 338 (2001)
9. Fernandez-Garcia, M., Martinez-Arias, A., Hanson, J.C., Rodriguez, J.A.: Nanostructured oxides in chemistry: Characterization and properties. Chem. Rev. **104**, 4063 (2004)
10. Buhro, W.E., Colvin, V.L.: Semiconductor nanocrystals: Shape matters. Nat. Mater. **2**, 138 (2003)
11. Burda, C., Chen, X.B., Narayanan, R., El-Sayed, M.A.: Chemistry and properties of nanocrystals of different shapes. Chem. Rev. **105**, 1025 (2005)
12. O'Regan, B., Gratzel, M.: A low-cost, high-efficiency solar cell based on dye-sensitized colloidal TiO_2 films. Nature **353**, 737 (1991)
13. Li, W., Osora, H., Otero, L., Duncan, D.C., Fox, M.A.: Photoelectrochemistry of a substituted-$Ru(bpy)_3^{2+}$-labeled polyimide and nanocrystalline SnO_2 composite formulated as a thin-film electrode. J. Phys. Chem. A **102**, 5333 (1998)
14. Bedja, I., Kamat, P.V., Hua, X., Lappin, A.G., Hotchandani, S.: Photosensitization of nanocrystalline ZnO films by Bis(2,2'-bipyridine)(2,2'-bipyridine-4,4'-dicarboxylic acid) ruthenium(II). Langmuir **13**, 2398 (1997)

15. Kavan, L., Kratochvilova, K., Gratzel, M.: Study of nanocrystalline TiO_2 (anatase) electrode in the accumulation regime. J. Electroanal. Chem. **394**, 93 (1995)

16. Croce, F., Appetecchi, G.B., Persi, L., Scrosati, B.: Nanocomposite polymer electrolytes for lithium batteries. Nature **394**, 456 (1998)

17. Bach, U., Lupo, D., Comte, P., Moser, J.E., Weissortel, F., Salbeck, J., Spreitzer, H., Gratzel, M.: Solid-state dye-sensitized mesoporous TiO_2 solar cells with high photon-to-electron conversion efficiencies. Nature **395**, 583 (1998)

18. Hagfeldt, A., Gratzel, M.: Light-induced redox reactions in nanocrystalline systems. Chem. Rev. **95**, 49 (1995)

19. Yamazoe, N.: New approaches for improving semiconductor gas sensors. Sens. Actuators B **5**, 7 (1991)

20. Weller, D., Moser, A.: Thermal effect limits in ultrahigh-density magnetic recording. IEEE Trans. Magn. **35**, 4423 (1999)

21. Sun, S.H., Murray, C.B., Weller, D., Folks, L., Moser, A.: Monodisperse FePt nanoparticles and ferromagnetic FePt nanocrystal superlattices. Science **287**, 1989 (2000)

22. Dai, J., Tang, J.K., Hsu, S.T., Pan, W.: Magnetic nanostructures and materials for magnetic random access memory. J. Nanosci. Nanotechnol. **2**, 281 (2002)

23. Chen, M., Nikles, D.E.: Synthesis, self-assembly, and magnetic properties of $Fe_xCo_yPt_{100-x-y}$ nanoparticles. Nano Lett. **2**, 211 (2002)

24. Colvin, V.L., Schlamp, M.C., Alivisatos, A.P.: Light-emitting diodes made from cadmium selenide nanocrystals and a semiconducting polymer. Nature **370**, 354 (1994)

25. Feldmann, C.: Preparation of nanoscale pigment particles. Adv. Mater. **13**, 1301 (2001)

26. Kukushkin, S.A., Slyozov, V.V.: Crystallization of binary melts and decay of supersaturated solid solutions at the ostwald ripening stage under non-isothermal conditions. J. Phys. Chem. Solids **56**, 1259 (1995)

27. Mutaftschiev, B.: Nucleation. In: Handbook of Crystal Growth, p. 187. Elsevier, Amsterdam (1993)

28. Paul, A.: Chemistry of glasses. Chapman & Hall, New York (1982)

29. Navarro, J.M.F.: El Vidrio, 2nd edn. C.S.I.C., Madrid (1991)

30. Fokin, V.M., Zanotto, E.D., Yuritsyn, N.S., Schmelzer, J.W.: Homogeneous crystal nucleation in silicate glasses: A 40 years perspective. J. Non-Cryst. Solids **352**, 2681 (2006)

31. Kingery, W.D., Berg, M.: Study of the initial stages of sintering solids by viscous flow, evaporation-condensation, and self-diffusion. J. Appl. Phys. **26**, 1205 (1955)

32. Kingery, W.D., Narasimhan, M.D.: Densification during sintering in the presence of a liquid Phase. II. Experimental. J. Appl. Phys. **30**, 307 (1959)

33. Kingery, W.D.: Densification during sintering in the presence of a liquid phase. I. Theory. J. Appl. Phys. **30**, 301 (1959)

34. Kingery, W.D., Bowen, H.K., Uhlmann, D.R.: Introduction to Ceramics, 2nd edn. Wiley, New York (1976)

35. Chiang, Y.M., Birnie, D.P., Kingery, W.D.: Physical Ceramics. Wiley, New York (1997)

36. Yin, Y., Alivisatos, A.P.: Colloidal nanocrystal synthesis and the organic-inorganic interface. Nature **437**, 664–670 (2005)

37. Jun, I.-W., Choi, J.-S., Cheon, J.: Shape control of semiconductor and metal oxide nanocrystals through nonhydrolytic colloidal routes. Angew. Chem. Int. Ed Engl. **45**, 3414–3439 (2006)

38. Lamer, V.K., Dinegar, R.H.: Theory, production and mechanism of formation of monodispersed hydrosols. J. Am. Chem. Soc. **72**, 4847–4854 (1950)

39. Turkevic, J., Kim, G.: Palladium – preparation and catalytic properties of particles of uniform size. Science **169**, 873–879 (1970)

40. Leite, E.R.: Nanocrystals assembled from the bottom up. In: Nalwa H.S (ed.) Encyclopedia of Nanoscience and Nanotechnology, vol. 6, pp. 537–554. American Scientific Publishers, Stevenson Ranch (2004)

41. Allen, S.M., Thomas, E.L.: The structure of materials, p. 447. Wiley, New York (1999)

42. Herring, C.: Some theorems on the free energies of crystal surfaces. Phys. Rev. **82**, 87–93 (1951)
43. Polleux, J., Pinna, N., Antonietti, M., Niederberger, M.: Ligand-directed assembly of preformed titania nanocrystals into highly anisotropic nanostructures. Adv. Mater. **16**, 436–439 (2004)
44. Colfen, H., Antonietti, M.: Mesocrystals and nonclassical crystallization, p. 228. Wiley, Hoboken (2008)
45. Ribeiro, C., Lee, E.J.H., Longo, E., Leite, E.R.: A kinetic model to describe nanocrystal growth by the oriented attachment mechanism. Chemphyschem **6**, 690–696 (2005)
46. Zheng, H.M., Smith, R.K., Jun, Y.W., Kisielowski, C., Dahmen, U., Alivisatos, A.P.: Observation of single colloidal platinum nanocrystal growth trajectories. Science **324**, 1309–1312 (2009)

Chapter 2
Basic Principles: Thermodynamics and Colloidal Chemistry

To obtain a complete understanding of the crystallization and growth process of a nanocrystal in a colloidal dispersion, knowledge of basic thermodynamics and colloidal chemistry principles is essential. For instance, the classical crystallization process is based on a Gibbs free energy (G) analysis where the free energy of the bulk crystal is considered as a surface free energy contribution. Thus, a brief introduction of thermodynamics principles is important to facilitate an analysis of the crystallization and growth process. Therefore, a revision of the basic principles related to colloidal chemistry will be presented because the main bottom-up chemical approach to synthesize nanocrystals is based on a colloidal synthesis process [1, 2].

2.1 Thermodynamics Basic Principles

Thermodynamics is one of the fundamental pillars of physical chemistry and material science and is essentially related to the conservation of energy and energy transference to predict the spontaneous direction of the chemical process and its equilibrium state. Classical thermodynamics is based on four laws the Zeroth law involves temperature definition and thermal equilibrium, and the first law is related to the conservation of energy and is mathematically described as

$$\Delta U = w + q \tag{2.1}$$

where ΔU is the internal energy variation from a final position to an initial position, w is the work, and q is the heat.

The second law is the spontaneous direction of processes as determined by the entropy (S) variation. Finally, the third law deals with entropy and states that a perfect solid has the same entropy at temperature (absolute scale) $T = 0$; this value may be taken to be $S = 0$.

E.R. Leite and C. Ribeiro, *Crystallization and Growth of Colloidal Nanocrystals*, SpringerBriefs in Materials, DOI 10.1007/978-1-4614-1308-0_2, © Edson Roberto Leite and Caue Ribeiro 2012

The first law is conservative and cannot be used to predict the spontaneous direction of the chemical process and its equilibrium state, so the second law accomplishes this prediction by considering the total entropy, i.e., the entropy of the system plus the entropy of its surroundings. If the total entropy increases, the process will be spontaneous. The entropy changes of the system and its surroundings can be incorporated into a new thermodynamics function of the system (the free energy).

The Gibbs and Helmholtz free energies arise from the Clausius inequality ($dS_{syst} + dS_{sur} \geq 0$). At a constant pressure (p) (a convenient experimental condition), we can define the Gibbs free energy (G) as

$$G = H - TS \tag{2.2}$$

where H is the enthalpy, T is the absolute temperature (K), and S is the entropy. This thermodynamic function takes into account an energetic parameter (H) and a structural parameter (S). Differentiating Eq. 2.2 and considering a constant temperature, this expression becomes:

$$dG = dH - TdS \tag{2.3}$$

Considering the Gibbs free energy, the spontaneous process will occur when $dG < 0$, and the equilibrium condition is given by $dG = 0$.

Combining the first law, the enthalpy definition ($H = U + pV$, where V is the volume) and the second law of thermodynamics, Eq. 2.3 is rewritten as

$$dG = Vdp - SdT \tag{2.4}$$

This very important equation facilitates an analysis of how G varies with important experimental parameters such as temperature (T) and pressure (p).

Considering an ideal gas and constant temperature, Eq. 2.4 is rewritten as

$$dG = nRT(dp/p) \tag{2.5}$$

where n is the number of moles and R is the gas constant.

Integrating from p_A to p_B gives

$$\Delta G = (G_B - G_A) = nRT \cdot \ln(p_B/p_A) \tag{2.6}$$

If we define $p_A = p^\phi = 1$ bar, Eq. 2.6 becomes (writing G_B as G)

$$G_m = G_m{}^\phi + RT \cdot \ln(p/p^\phi) \tag{2.7}$$

where $G_m = G/n$ is the molar Gibbs free energy for a pure substance and is equal to the chemical potential μ, which for an ideal gas becomes

$$\mu = \mu^\phi + RT \cdot \ln(p/p)^\phi \tag{2.8}$$

where μ^ϕ is the standard chemical potential.

Table 2.1 List of the different forms to describe the activities

Activity	System	Definition
Partial pressure	Ideal gas	$a_j = (p_j/p^{\phi})$
Fugacity (f_j)	Real gas	$a_j = (f_j/p^{\phi})$
Activity (a_j)	Solvent/solute	$a_j = \xi_j \cdot x_j$; x_j is the mole fraction, and ξ_j activity coefficient

The definition of chemical potential for a pure substance is

$$\mu = (\partial G/\partial n)_{p,T} \tag{2.9}$$

Considering a mixture of different substances, this definition is modified to

$$\mu_j = (\partial G/\partial n)_{p,T,nj} \tag{2.10}$$

where nj is the number of moles of species j.

Analyzing Eq. 2.10, we can conclude that the Gibbs free energy will depend on the chemical composition nj as well as T and p. Thus, we can modify Eq. 2.4 as:

$$dG = Vdp - SdT + \sum \mu_j dn_j \tag{2.11}$$

This equation is known as the fundamental equation of chemical thermodynamics. Other terms are added to Eq. 2.11 to provide a general description of how different parameters can modify the Gibbs free energy and the spontaneous direction of the chemical process.

The chemical potential is an important thermodynamic parameter; however, because it was derived by considering an ideal gas (i.e., a system without inter-molecular interactions), Eq. 2.8 must be rewritten to consider activities (a_j) for the chemical potential. The activities are an effective concentration which take into account the interaction among molecules in gas and liquid state. Table 2.1 lists the different forms to describe these activities.

Considering the activity, the chemical potential for a solvent/solute system can be rewritten as

$$\mu_j = \mu^* + RT \cdot \ln x_j + RT \ln \xi_j \tag{2.12}$$

where μ_j is the chemical potential of the component j and μ^* is the chemical potential of the pure solvent.

In materials chemistry, an important parameter to be added to the fundamental equation of chemical thermodynamics is the surface tension or surface energy (γ) which in a liquid is defined as the reversible work (w) required to increase the surface of the liquid by a unit area (A):

$$dw = \gamma dA \tag{2.13}$$

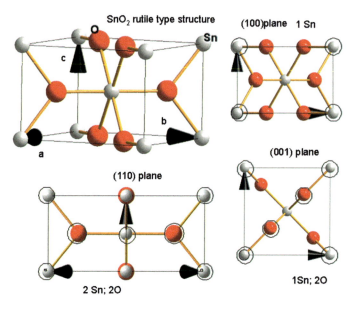

Fig. 2.1 Tin oxide (SnO_2) crystallographic structure; (*hkl*) planes present different chemical compositions. In the plane figures, the hollow circle shows atoms in the same plane

Under specific conditions, this surface energy can be related to other thermodynamic properties of the system which facilitates rewriting Eq. 2.11 as

$$dG = Vdp - SdT + \sum \mu_j dn_j + \gamma dA. \tag{2.14}$$

In a solid, γ is defined as the reversible work performed to create a new surface by adding additional atoms to the surface [3]. For a finite change in the surface area at constant p, T, and nj, ΔG is given by

$$\Delta G = \int \gamma dA. \tag{2.15}$$

In a crystalline solid, the new surface must be created without changing the crystallographic orientation of the new surface in relation to the existing surface. In addition, in a crystalline solid, γ is not isotropic and depends on the crystallographic orientation which is especially critical in nonmetallic solids where crystallographic planes show different chemical compositions. For SnO_2, Fig. 2.1 displays different chemical compositions of three crystallographic (*hkl*) planes. For instance, the (100) plane is composed of tin (Sn) only while the (001) plane is formed by atoms of Sn and O (oxygen) in a ratio of 1:2.

An important consequence of the nonisotropic nature of the γ in a crystalline solid is the shape that a single crystal will achieve under equilibrium conditions. For instance, if the surface energy per unit area of a single-crystalline particle varies with the crystallographic plane (*hkl*) of the surface, then the shape of the crystal that minimizes its surface energy for a given volume will not be spherical; this tendency can lead to faceting of crystal surfaces.

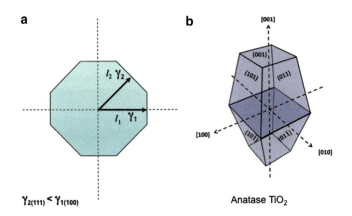

Fig. 2.2 (**a**) Wulff shape for a hypothetical bidimensional crystal. In this figure, $\gamma_1 = \gamma_{(100)}$ and $\gamma_2 = \gamma_{(111)}$. (**b**) Wulff shape of a TiO$_2$ crystal (with tetragonal anatase structure) based on ab initio calculated surface energy

Independently, Giggs and Curie [4], in an independent way, defined the equilibrium shape of a crystal as the possession of a minimal total surface energy for a constant volume. For a given crystal, this criterion can be described as

$$\Sigma \gamma_i S_i = \text{Minimum (constant volume)} \tag{2.16}$$

where γ_i is the specific free surface energy of the crystal ith face, and S_i is the facet i surface area.

Wulff [5, 6] shows that this criterion is equivalent to

$$\Sigma \gamma_i / h_i = \text{Constant} \tag{2.17}$$

where h_i is the central distance of the ith facet (the Wulff rule) which is basic for all theories related to crystal shapes. According to this rule, the equilibrium shape can be constructed by the following geometric procedure: From an arbitrary point in space, vectors normal to a possible crystallographic facet are drawn. Lengths proportional to the corresponding values γ_i are plotted on each vector, and planes perpendicular to the vector at their tips are drawn, thus forming a polyhedron that represents the equilibrium shape.

This procedure is illustrated in Fig. 2.2a, for a two-dimensional cubic crystal having $\gamma_{(111)} < \gamma_{(100)}$. The plot obtained (known as Wulff construction) indicates that the facets {111} will be larger than the facets {100}. The Wulff construction is an interesting way to predict the shape of the nanocrystal. However, the main problem concerning this construction is to obtain experimental data about the surface energy of different (*hkl*) planes. This problem can be overcome using theoretical data obtained from first principium quantum mechanical chemical calculation [7]. For instance, Fig. 2.2b shows the Wulff construction of a TiO$_2$ crystal (with tetragonal anatase structure) based on ab initio calculated surface energy [8, 9].

2.2 Colloidal Chemistry Basic Principles

A colloid is a dispersion of small particles (i.e., particles smaller than 1 μm) or a phase of one material in another material. Basically, a colloidal particle is formed by clusters of numerous atoms or molecules which are too small to be visible using an ordinary optical microscopy. A useful classification of colloids is lyophilic (solvent-attracting) and lyophobic (solvent-repelling). If the solvent is water, the terms lyophilic and lyophobic are replaced by hydrophilic and hydrophobic, respectively.

In general, the disperse phase of a colloid is thermodynamically unstable with respect to the bulk. This instability can be expressed thermodynamically by Eq. 2.13. Considering a constant T, p, and composition (n), dG is equal to γdA. Therefore, dG will be smaller than zero ($dG < 0$) if $dA < 0$. Thus, to decrease the interface area and γ, the disperse phase will form large aggregates which promote colloidal coagulation and a type of phase segregation. This analysis shows that colloids are thermodynamically unstable, although they can be kinetically stabilized.

Three different colloidal states can be defined (see Fig. 2.3). In the disperse state, particles (or phases) in the suspension repel other particles. In this state, the repulsive interaction is the dominant potential energy among the particles. Considering the weakly and strongly flocculated states, the attractive interaction dominates the potential energy, and the difference between those states is the volume fraction of the disperse phase. In the weakly flocculated state, particles aggregate in clusters

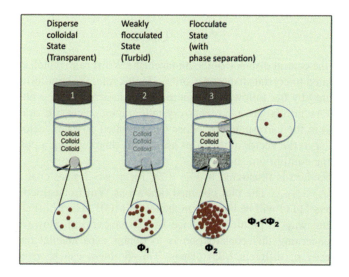

Fig. 2.3 Schematic representation of the relationship between the visual aspect of the colloidal state and the resulting suspension structure. In this diagram, vial 1 shows a weakly flocculated colloidal state, vial 2 shows a disperse colloidal state, and vial 3 shows a strongly flocculated state

Table 2.2 Common types of interaction forces and potential energy among atoms, ions, and molecules [13]

Type of interaction	Interparticle distance dependence of the potential energy	Temperature dependence of the potential energy
Dipole-charge	$\propto L^{-4}$	$\propto T^{-1}$
Dipole-dipole (Keeson)	$\propto L^{-6}$	$\propto T^{-1}$
Dipole-induced-dipole (Debye)	$\propto L^{-6}$	Not dependent
Induced dipole-induced-dipole (LD)	$\propto L^{-6}$	$\propto T$

or agglomerate in a suspension at a volume fraction of particles (ϕ) below the gel point. In this case, an equilibrium separation distance among agglomerates can be defined. In the strongly flocculate state, the ϕ is higher than the gel point, resulting in a touching particle network where a strong sedimentation and phase separation is observed.

Considering the previous discussion, careful control of the interparticle forces can result in the preparation of colloidal suspensions in one of the described states. Thus, an analysis of these forces facilitates control of the colloidal state.

Colloidal stability is controlled by the total interparticle potential energy (V_T) which can be described as

$$V_T = V_{vdW} + V_e + V_{steric} + V_{structural} \qquad (2.18)$$

where V_{vdW} is the attractive potential energy due to long-range van der Waals (vdW) interaction between particles, V_e is the repulsive potential energy resulting from electrostatic interaction between charged particles, V_{steric} is the repulsive potential energy resulting from steric interaction between particle surfaces coated with an adsorbed macromolecule species, and $V_{structural}$ is the potential energy related to the presence of nonadsorbed species in solution that can modify the suspension stability. The first two terms of Eq. 2.18 constitute the DLVO theory developed by Derjaguin and Landau [10] and Verwey and Overbeek [11].

vdW forces result from three or more additive terms: the Keeson force, the Debye force, the dipole-charge interaction force, and London dispersion (LD) forces. Table 2.2 lists common types of interaction forces among atoms, ions, and molecules and its potential energy dependence regarding the interparticle distance (L) and temperature (T) [12].

In some molecules as well as in ionic and covalent crystals, the atomic arrangement of ions or atoms can induce a permanent dipole. For crystals, this dipole can be associated with a specific crystallographic orientation. For instance, permanent dipoles, charges, or external electrical fields can also induce dipoles. These dipoles can interact not only with one another but also with charges, particularly in aqueous solutions. It is not a simple task to analyze dipole-dipole and dipole-charge interactions; moreover, in solution, thermal fluctuations must be considered.

However, the dipole-charge $(\langle V \rangle_{\text{d-c}})$ and dipole-dipole $(\langle V \rangle_{\text{d-p}})$ potentials (Keeson) can be described in a simplified way as

$$\langle V \rangle_{\text{d-c}} = -c^2 q^2 p^2 / 3k_b TL^4, \tag{2.19}$$

$$\langle V \rangle_{\text{d-p}} = -2/3(c^2 p1^2 p2^2 / 3k_b TL^6), \tag{2.20}$$

where $c = 1/4\pi\varepsilon_0$ (ε_0 is the permittivity of the vacuum, 8.85×10^{-12} C²/Nm²), q is the charge, p is the dipole moment, T is the temperature (K), and L is the distance between molecules. Both equations consider a Boltzmann distribution to determine the value of the mean potential energy ($\langle V \rangle$), considering all possible angular orientations between the dipole-charge and the dipole-dipole. For these two equations, dipole-charge interactions fall off as $1/L^4$ (medium-range interactions) while dipole-dipole interactions have short-range interaction potentials (falling off as $1/L^6$). Both potentials show a dependence of $1/T$ in relation to temperature. Debye interactions occur between dipole-induced dipoles which present a dependence on $1/L^6$ and are independent of temperature. The LD interaction of two induced dipoles also obeys a power-law relationship with an L^{-6} power for potential energy. The temperature dependence of LD interactions is given by the Hamaker constant which is directly proportional to T [13, 14].

For spherical particles of the same dimension and considering the predominance of dispersion forces, V_{vW} is given by

$$V_{\text{vW}} = -A/6[(2/s^2 - 4) + (2/s^2) + \ln(s^2 - 4/s^2) \tag{2.21}$$

where $s = (2a + h)/a$, h is the minimum separation between particles of radius a, and A is the Hamaker constant. The Hamaker constant depends on the nature of the disperse phase (or particle) as well as the medium. For instance, the Hamaker constant for SiO_2 (the quartz crystalline phase) is 8.86×10^{-20} J in vacuum and 1.02×10^{-20} J in water. For amorphous SiO_2, the values are 6.5×10^{-20} J in vacuum and 0.46×10^{-20} J in water [15].

The repulsive V_e shows exponential distance dependence. The strength of this interaction depends on the surface potential induced by the interaction between the particle and the medium. For spherical particles of similar size (radius a), the interaction V_e, considering constant potential, is

$$V_e = 2\pi\varepsilon_r\varepsilon_o\Psi_o^2 \ln[1 + \exp(-\kappa h)], \tag{2.22}$$

where ε_r is the dielectric constant of the medium, ε_o is the vacuum permittivity, Ψ_o is the surface potential, and $1/\kappa$ is the Debye-Huckel screening length. κ is given by:

$$\kappa = [(F^2\Sigma N_i Z_i^2)/(\varepsilon_r\varepsilon_o kT)]^{1/2}, \tag{2.23}$$

where N_i and z_i are the number density and valence of type I counterions, k is the Boltzmann constant, and F is the Faraday constant.

One important aspect of V_e is the physical chemistry control it provides over the surface potential which facilitates the modification of colloid stability in a charged system (mainly in water as the solvent). For instance, the surface charge of metal oxide can be controlled by the protonation or deprotonation of the hydrous oxide particle surface (M−OH). The ease with which protonation and deprotonation occurs depends on the metal atoms and can be controlled by the pH. The pH at which the particles present a net zero charge is called isoelectric point. At a pH > isoelectric point, the particles are negatively charged while at the pH < isoelectric point, the particles are positively charged. Equations 2.24 and 2.25 show the effect of pH in surface particle charges:

$$M - OH + OH^- \rightarrow M - O^- + H_2O \qquad (2.24)$$

$$M - OH + H^+ \rightarrow M - OH_2{}^+ \qquad (2.25)$$

The surface charge of a disperse particle in a colloidal dispersion can be determined by a zeta potential (ζ) measurement, i.e., the potential where the ions are less firmly bound to the particle surface (diffuse layer). Taking the surface potential to be equal to the zeta potential, direct analyses of V_e can be obtained by analyzing ζ. In fact, the magnitude of ζ can be taken as a parameter to evaluate colloidal stability. As a rule, particles with $\zeta \geq \pm 30$ mV can be considered stable. Figure 2.4a shows a typical plot of zeta potential as a function of pH for a SnO_2 nanocrystalline colloidal dispersion in water where the isoelectric point is approximately pH $= 3.0$. This plot provides supplementary information such as the colloidal stability pH region (green region) where the disperse state dominates as well as the pH region where weak and strong flocculation occurs (orange region). There is a third region (blue) where the particles present high zeta potential; however, any small variation in the pH provokes a considerable variation in ζ. This sensitive pH region should be avoided for a colloidal suspension with high stability. Figure 2.4b is a visual analysis of the colloidal suspension in the disperse state and in the weakly flocculated state. In this picture, the difference between the vials is the pH, and the solvent (water) and particle concentration are the same. The colloidal dispersion in disperse state presents a pH $= 9.5$, and the colloidal dispersion in a weakly flocculated state presents a pH $= 4.3$.

Considering Eq. 2.18 again, the last two terms are directly related to the conformational structure of the molecules or macromolecules attached to the particle surface (V_{steric}) as well as to the interaction between the solvent and the disperse phase ($V_{structural}$). If those terms are the dominant in the total potential, the entropy of the system will be important, and a colloid can be thermodynamically stabilized which will happen when the disperse phase behaves as in a solution such as in polymer solutions and soluble nanocrystals. Equation 2.12 demonstrates that the component j chemical potential of the mixture will be smaller than the pure solvent chemical potential, since the terms $RT \cdot \ln x_j + RT \cdot \ln \xi_j$ are <0 because x_j and ξ_j are < 0.

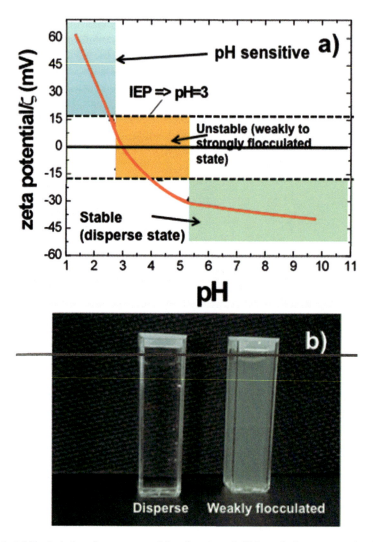

Fig. 2.4 (**a**) Typical plot of a zeta potential as function of pH for a SnO$_2$ nanocrystal colloidal dispersion in water. (**b**) Visual analysis of the colloid in the disperse state and in the weakly flocculated state

Steric stabilization provides an interesting route to control colloidal stability and is widely used to control the particle size and the colloidal stability of nanocrystals during the synthesis process, especially in organic solvents [1, 2]. In this route, adsorbed organic molecules are utilized to induce steric repulsion. To be effective, the adsorbed layer must be of sufficient thickness and density to overcome the V_{vw} attraction between particles and to prevent bridging flocculation. The conformation of the organic layer can vary dramatically, depending on the interaction with the solvent, the molecule weight, architecture, and colloidal and organic concentration.

This kind of approach can modify the colloidal affinity to a different solvent. For instance, the attachment of a low-polarity organic molecule to the surface of an oxide dispersed in a polar solvent can transform the disperse phase to become lyophobic which induces flocculation and phase segregation. The attachment of organic molecules to the surface of a nanocrystal is an important issue because the assemblies as well as the chemical functionality of these nanocrystals depend basically on their ability to attach organic molecules with different structures and polarities in the nanocrystal surface. Actually, nanocrystals can be developed that present amphiphilic characteristics, i.e., nanocrystals with simultaneous lyophobic and lyophilic characteristics [16].

References

1. Yin, Y., Alivisatos, A.P.: Colloidal nanocrystal synthesis and the organic-inorganic interface. Nature **437**, 664–670 (2005)
2. Jun, I.-W., Choi, J.-S., Cheon, J.: Shape control of semiconductor and metal oxide nanocrystals through nonhydrolytic colloidal routes. Angew. Chem. Int. Ed. **45**, 3414–3439 (2006)
3. Allen, S.M., Thomas, E.L.: The Structure of Materials. Wiley, New York (1997)
4. Givargizov, E.I.: Oriented Crystallization on Amorphous Substrates. Plenum Press, New York/London (1991)
5. Wulff, G.: On the question of speed of growth and dissolution of crystal surfaces. Z. Kristallogr. **34**, 449–530 (1901)
6. Herring, C.: Some theorems on the free energies of crystal surfaces. Phys. Rev. **82**, 87–93 (1951)
7. Stroppa, D.G., Montoro, L.A., Beltran, A., Conti, T.G., da Silva, R.O., Andres, J., Longo, E., Leite, E.R., Ramirez, A.J.: Unveiling the chemical and morphological features of $Sb-SnO_2$ nanocrystals by the combined use of high-resolution transmission electron microscopy and ab initio surface energy calculations. J. Am. Chem. Soc. **131**, 14544–14548 (2009)
8. Lazzeri, M., Vittadini, A., Selloni, A.: Structure and energetics of stoichiometric TiO_2 anatase surfaces. Phys. Rev. B: Condens. Matter Mater. Phys. **63**, 155409 (2001)
9. Diebold, U.: The surface science of titanium dioxide. Surf. Sci. Rep. **48**, 53–229 (2003)
10. Derjaguin, B.V., Landau, L.D.: Theory of the stability of strongly charged lyophobic sols and of the adhesion of strongly charged particles in solution of electrolytes. Acta Physicochim. URSS **14**, 633–652 (1941)
11. Verwey, E.J.W., Overbeek, J.Th.G.: Theory of Stability of Lyophobic Colloids. Elsevier, Amsterdam (1948)
12. Israelachvili, J.N.: Intermolecular and Surface Forces. Academic, London (1992)
13. Parsegia, Va, Ninham, B.W.: Temperature-dependent van der Waals forces. Biophys. J. **10**, 664–673 (1970)
14. Bergstrom, L.: Hamaker constants of inorganic materials. Adv. Colloid Interface Sci. **70**, 125–169 (1997)
15. Lewis, J.A.: Colloidal processing of ceramics. J. Am. Ceram. Soc. **83**, 2341–2359 (2000)
16. Gonçalves, R.H., Cardoso, C.A., Leite, E.R.: Synthesis of colloidal magnetitenanocrystals using high molecular weight solvent. J. Mater. Chem. **20**, 1167–1172 (2010)

Chapter 3
Classical Crystallization Model: Nucleation and Growth

The nucleation and growth process is a well-accepted model to describe several processes involving the crystallization of a condensed phase. This model can be applied to describe the crystallization process of single elements such as metal, where the liquid phase and the crystalline phase present the same chemical composition, as well as the crystallization of covalent and ionic crystals processed by wet chemical route. In this case, the process is much more complicated due to the presence of different chemical composition and, consequently, a chemical potential between the solid phase and the liquid phase.

Many of the synthesis methods to produce nanoparticles are based on precipitation steps or nucleation and growth in the reaction media [1–3]. Precipitation reactions involve the simultaneous occurrence of these steps, as well as coarsening and agglomeration processes [4–6]. Due to the difficulties in isolating each process for independent study, the fundamental mechanisms of precipitation are still not entirely understood. However, a good understanding of the nucleation step is fundamental for grasping the nature of nanosize particles, since it will be determinant in the following steps.

An ideal condition of synthesis is presented in the graphic of Fig. 3.1a, where we can observe that the nucleation rate curve as function of a given experimental parameter (such as temperature, concentration, and pH) is not overlapping the growth rate. This condition is ideal because we can control the number of nucleus without any growth process. During the overlapping between nucleation and growth process (Fig. 3.1b), the events will not be independents. As a consequence, we will observe a poor control over the size, size distribution, and morphological and colloidal stability of the crystalline phase.

In the next sections, a didactic division is adopted, to help the reader in understanding the interrelation of the topics with the nanocrystals synthesis.

E.R. Leite and C. Ribeiro, *Crystallization and Growth of Colloidal Nanocrystals*, SpringerBriefs in Materials, DOI 10.1007/978-1-4614-1308-0_3,

Fig. 3.1 (**a**) Independent nucleation and growth process showing the nonexistence of overlapping between the nucleation and growth rate as function of an experimental parameter. (**b**) Dependent nucleation and growth process as function of an experimental parameter. In this case, the overlapping between the nucleation and growth rate is observed

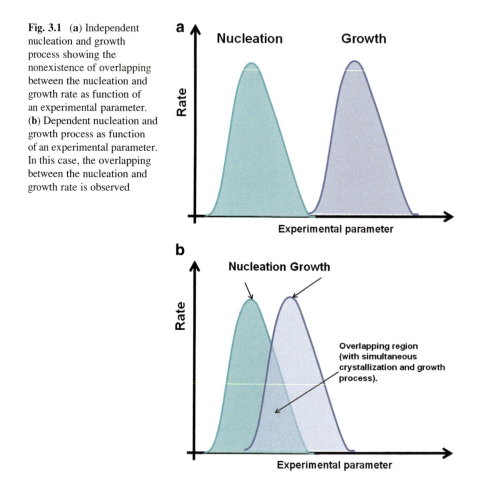

3.1   The Bubble Model

The formation of a solid in a homogeneous system, like melts or reactional medium, is strictly the problem of obtaining a heterogeneous phase – a solid – into the homogeneous phase. In the homogeneous system, all the atoms or molecules have high freedom degrees, and the order is minimal, despite some organization may be observed [7]. In a very simple point of view, the formation of the heterogeneous phase is similar to the stabilization of a bubble in a liquid, as illustrated in Fig. 3.2.

The analysis of the problem stems in the balance of the driving force for the bubble formation and the total work W_{sup} to obtain a new surface, i.e.:

$$W_{sup} = \gamma dA, \tag{3.1}$$

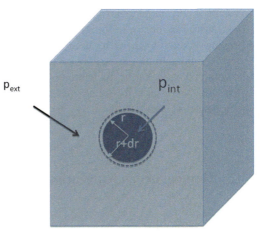

Fig. 3.2 Scheme of a bubble in a liquid, showing the distribution of forces over the bubble

where γ is the surface energy per unit area of the bubble. Since the most stable geometric condition is the sphere, one can relate the area variation to the total sphere area, i.e., $dA = d(4\pi r^2) = 8\pi r dr$, where r is the sphere radius. Then, substituting the terms in Eq. 3.1, one has:

$$W_{sup} = 8\pi \gamma r dr, \qquad (3.2)$$

where the term $8\pi \gamma r$ is easily identified as a force term, related to the bubble formation. In the equilibrium, the bubble is limited by two pressures, the internal p_{int} and external p_{ext} pressures (commonly atmospheric pressure). To assure the static condition, the internal force $p_{int} \cdot A_{sphere}$ needs to be equal to the external force $p_{ext} \cdot A_{sphere} + 8\pi \gamma r$. Comparing the terms, one can deduce the expression:

$$p_{int} - p_{ext} = p_{eff} = 2\gamma / r. \qquad (3.3)$$

The expression is known as Laplace-Young equation, and this simple relation shows that the stability of a heterogeneous phase depends on the surface energy of the phase. The idea was explored in nanocrystals considering the unbalanced forces in small particles as pressure excess upon the whole solid particle [3, 8, 9]. Zhang and Banfield [6] studied the transformation of anatase to rutile in TiO_2 nanocrystals of 5–100 nm, observing different temperatures to the transformation yield. They assumed that the increase in activation energy would be proportional to the excess pressure, $E_a = E_{ac} + C \cdot p_{eff} = E_{ac} + C'/r$, where E_a is the effective activation energy, E_{ac} is the bulk energy activation, and C, C' are proportionality constants. The difference in E_a to the total range was around 60 kJ/mol, a significant value in view of the estimated bulk activation energy ($E_{ac} = 185$ kJ/mol).

Tolbert and Alivisatos [10–12] investigated the pressure-induced transformation of wurtzite to rocksalt in CdSe nanocrystals. The authors showed that a significant increase in the pressure necessary to induce the transformation followed a scaling law of the type $p_{transf} \cong 1 + C''/r$. They observed an increase of 35% in transition pressure in 10-nm nanoparticles in relation to 21-nm nanoparticles (3.6–4.9 GPa). The transformations were found to be fully reversible, albeit with some hysteresis, showing an energy barrier to direct transformation.

This approach, despite it may be understood only as a physical analogy, helps the comprehension of the nucleation process, since the main idea remains the same: what are the criteria to define the stability of a heterogeneous phase into the homogeneous phase?

3.2 Homogeneous Nucleation

The formation of a nucleus in a homogeneous system is not trivial, since in this case one can suppose that the event is entirely statistical, i.e., the nucleus will be formed due to unbalanced forces or variations in local stoichiometry [2, 4, 13–17]. The simplest way is to consider the stability of a given nucleus or small cluster as given by the balance between the free energy of phase formation, ΔG_V, and the work given by the new surface, ΔA (the product of the surface free energy and the surface area), as follows for a sphere [2, 18–20]

$$\Delta G = -4/3\pi r^3 \cdot \Delta G_V + 4\pi r^2 \gamma, \tag{3.4}$$

where ΔG is the free Gibbs energy associated to the process. Making the first derivative of the expression equal to zero, $d(\Delta G)/dr = 0$, one has the critical radius for a stable nucleus (r_{crit}):

$$r_{crit} = 2\gamma/\Delta G_V. \tag{3.5}$$

Returning the value for r_{crit} obtained in Eq. 3.5 in Eq. 3.4, the critical free energy may be defined as:

$$\Delta G_{crit} = 16 \cdot \pi\gamma^3/3 \cdot \Delta G_V{}^2 \tag{3.6}$$

The importance of the relation exposed here, remarkably similar to Eq. 3.3, is clear in a schematic plot, as shown in Fig. 3.3: the value r_{crit} defines the critical size for the survival of the nucleus in the media. Any nucleus formed below this size will be dissolved in the medium, and nucleus with size over that will grow. The analysis does not imply that nuclei with size equal to r_{crit} will be the minor particle size in the nucleated system, since in this case the free energy is not favorable yet, but it only implies that this nuclei will survive.

A practical conclusion from the equations is that each system will show a particular critical size, determined by energetic considerations. Therefore, the first problem is determining the surface energy of the desired system. Calorimetric measurements are a way to determine this property [21–28]. At atmospheric pressure, the effect of the volume variation, $P{\cdot}\Delta V$, is expected to be small, and

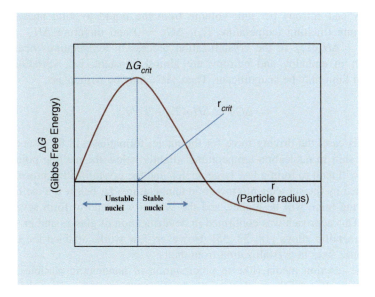

Fig. 3.3 Schematic plot of the Eq. 3.4

Table 3.1 Surface enthalpies and transformation enthalpies relative to bulk stable polymorph for oxides (Adapted from ref. [29])

Oxide	Surface enthalpy (J/m^2)
$\alpha - Al_2O_3$	2.6 ± 0.2
$\gamma - Al_2O_3$	1.7 ± 0.1
TiO_2 rutile	2.2 ± 0.2
TiO_2 anatase	0.4 ± 0.1
ZrO_2 monoclinic	6.5 ± 0.2
ZrO_2 tetragonal	2.1 ± 0.05

the surface entropy is also expected to be slight. Based on this assumption, the surface free energy can be approximated to the surface enthalpy (the variable that is properly measured in calorimetry). Table 3.1 shows surface enthalpies and transformation enthalpy data for some oxides and their polymorphs. An initial analysis of the data reveals that the surface enthalpy (or energy) decreases as the phase becomes more metastable (higher transformation enthalpy relative to the bulk stable polymorph), i.e., smaller surface energies lead to lower barriers to stabilization [24, 29]. This analysis (albeit not at all correct) indicates that metastable phases tend to nucleate more easily than stable ones, a statement generally referred as the Ostwald step rule [29]. In fact, it is well known that crystallization generally follows a sequence of metastable phases before the most stable phase is attained.

The second problem is determining the physical meaning of the term ΔG_V. In the case of nucleation in melts, the variable may be related to the thermal shift in the solidification process [3, 7, 18–20, 30–34]. From the definition of Gibbs free energy, $\Delta G_V = \Delta H_V - T\Delta S_V$, where ΔH_V and ΔS_V are the differences between enthalpy and entropy per unit volume from the liquid to solid phases. At the equilibrium (melting temperature T_m), $\Delta G_V = 0$ and therefore $\Delta H_V = T_m \cdot \Delta S_V$ or $\Delta S_V = \Delta H_V / T_m$. In the crystallization process in temperatures near T_m, the variation in enthalpy and entropy are almost constant, i.e., variations below the solid limit may be insignificant. Then, ΔG_V may be written as:

$$\Delta G_V \cong \Delta H_V (T_m - T)/T_m. \tag{3.7}$$

In this case, the driving force for the nuclei formation will be essential to the thermal shift in nucleation temperatures slightly below the melting point and the system will tend to form a few large nuclei, and the crystallization process will be dominated by the subsequent growth; in temperatures far below the melting point, the driving force will be higher, and the system will tend to form several small nuclei. This approach was confirmed in devitrification of glasses and in solidification of metallic alloys [18, 32, 35]. However, few works attained the control of nanometric sizes in crystallization from melts.

In the reaction media (like in precipitation of nanometric particles), a more complete analysis of the nucleation can be done taking into account the chemical potential of the nucleus. In such a condition, the nucleus possesses free Gibbs energy, as follows [1, 2, 31]:

$$dG = \mu_0 dn + \Sigma_i \gamma_i dA_i, \tag{3.8}$$

where μ_0 is the chemical potential of the bulk nucleus, n is the number of moles, γ_i is the surface energy, and A_i is the area for each surface i. The equation can be rearranged to

$$\mu = \mu_0 + V_m \sum_i \gamma_i dA_i / dV, \tag{3.9}$$

since $dn = dV/V_m$ (where V_m is the molar volume). For a particle of any shape, the surface area A_i and the volume V can be written as generic equations for a given Z characteristic dimension as follows:

$$A_i = p_i Z^2, \tag{3.10}$$

$$V = q Z^3, \tag{3.11}$$

where p_i, q are geometric constants. Taking the derivative of the two expressions as Z, and applying it to a chain rule, one has:

$$(dA_i/dZ)/(dV/dZ) = dA_i/dV = 2A_i/3V. \tag{3.12}$$

At this point, the surface energy can be inserted as an average surface tension. Here, the contributions from edges and corners are not considered, although this contribution may be important (it will be discussed later). For the sake of simplicity, the expression can be rearranged as follows:

$$\gamma_A = \sum \gamma_i A_i / \sum A_i. \tag{3.13}$$

Inserting the Eqs. 3.12 and 3.13 into Eq. 3.9 and rearranging them, one has:

$$\mu = \mu_0 + \alpha_F 2 V_m \gamma_A / 3Z, \tag{3.14}$$

where α_F is a shape factor defined as $\sum p_i / q$.

This equation describes the chemical potential of the formed nucleus. The chemical potential of the substance μ' (or reaction product) in solution or melt is:

$$\mu' = \mu_0' + RT \ln a, \tag{3.15}$$

where R is the universal gas constant, a is the activity in solution, and μ_0' is the standard chemical potential in solution. One can assume that the standard chemical potential μ_0 in the particle is $\mu_0 = \mu_0' + RT \ln a_0$, where a_0 is the saturation activity. Comparing the equilibrium condition for precipitation, i.e., $\mu = \mu'$, and rearranging it, one has:

$$\ln a/a_0 = \alpha_F 2 V_m \gamma_A / 3ZRT. \tag{3.16}$$

This equation is remarkable, since it describes a general form to express the relation among surface energies, chemical potential, and dimension Z. For a spherical particle, $Z = r$ and $\alpha_F = 3$, Eq. 3.16 reduces to:

$$r = 2 V_m \gamma_A / (RT \ln a/a_0), \tag{3.17}$$

which is easily compared to Eq. 3.5. Observing the equation above, the value ΔG_V is interpreted in terms of the variation of the chemical potential of the ideal solution (ion or molecule) times the molar volume. For very dilute systems, the approximation of activity to the concentration, $a \sim c$, is usual and commonly accepted as valid.

The practical application of the theory lies in adequate correlations with synthesis variables. Kukushkin and Nemna [36] analyzed the problem of the homogeneous nucleation in the precipitation of poorly soluble bases and salts, evaluating the role of pH in the nucleation. The authors assumed ideal solutions where the supersaturation is a pH function, i.e., the precipitation occurs by a pH change. To a poorly soluble base $[M(OH)_p]^q$, it is assumed that the concentration of hydroxide species is small and has little effect on crystallization kinetics. This is a condition attained in diluted solutions. The precipitation will follow the reaction:

$$M_{aq}^{n+} + nOH^- \rightarrow M(OH)n^-. \tag{3.18}$$

Since the system is considered as poorly soluble, the reaction is assumed in a single way. By definition, the chemical potential of a infinite crystal is $\mu_{s,\infty} = \mu_{l+}^* + n\mu_{l-}^*$, where the subscripts $l+$, $l-$ designate, respectively, the ions M^{n+} and OH^- in solution, with concentrations c_+ and c_-. Then, the difference between the chemical potential of a supersaturated and a saturated solution follows as:

$$\Delta\mu = (\mu_{l+} + \mu_{l-}) - (\mu_{l+}^* + \mu_{l-}^*) = RT(\ln(c_+ \cdot c_-^n) - \ln(c_+^* \cdot c_-^{*n})), \quad (3.19)$$

or

$$\Delta\mu/RT = \ln c_+ \cdot c_-/S_P, \quad (3.20)$$

where the solubility product is defined as $S_P = c_+^* \cdot c_-^{*n}$. Also, considering diluted solutions, low values in $\Delta\mu$ are expected, implicating that the right hand side of the Eq. 3.20 may be expanded in a power series, considering only two terms:

$$\Delta\mu/RT = (c_+ \cdot c_-^n - S_P)/S_P, \quad (3.21)$$

since $c_- = K_w \cdot 10^{pH}$, where K_w is the water dissociation constant, 10^{-14}, and pH is the value at the precipitation moment, Eq. 3.21 can be simplified to an expression dependent only of c_+. However, the term $\Delta\mu$ is equal to $RT \cdot \ln(a/a_0)$ in Eq. 3.17. Then, substituting Eq. 3.20 into Eq. 3.17 leads to:

$$r_{crit} = 2 \cdot V_M \cdot S_P \cdot \gamma/K_w^n \cdot 10^{n \cdot pH} \cdot c_+ - S_P. \quad (3.22)$$

This relation shows the interpretation of pH change as the driving force to nucleation. However, the relation will be valid only when Eq. 3.18 is applied. In the case of amphoteric poorly soluble bases, the precipitate may dissolve in high pH values, and in this case, an equilibrium equation needs to be written as follows:

$$M_{aq}^{n+} \leftrightarrow MOH_{aq}^{(n-1)+} + \leftrightarrow ...M(OH)_n \downarrow ... \leftrightarrow [M(OH)_m]^{(m-n)-}. \quad (3.23)$$

To solve the problem, it is necessary to write a set of equations defining the occurrence of the intermediary species in Eq. 3.23. This is not a trivial objective, and the final set cannot be solved analytically. Using steady-state approximations for the intermediaries, Kukushkin and Nemna [36] proposed a general equation for the precipitate $M(OH)_n$:

$$\Delta\mu/RT = (n^n \cdot (c_n^*(pH))^{n+1} - S_P)/S_P, \quad (3.24)$$

where $c_n^*(pH)$ is the concentration of $M(OH)_n$ after attaining the ionic equilibrium in solution, but prior to the precipitation (a function of pH, since in this case the solubility of the oxide is pH dependent). The equation may be substituted into Eq. 3.17, in the same way.

A plot of the Eqs. 3.21 and 3.22 is shown in Fig. 3.4, related to the precipitation of $Ca(OH)_2$ (a poorly soluble base). Since the reaction is assumed as irreversible,

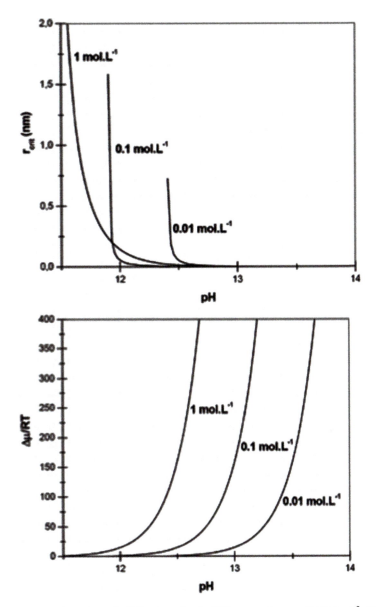

Fig. 3.4 Plot of driving force $\Delta\mu/RT$ and r_{crit} for Ca(OH)$_2$, assuming $\gamma = 0.066$ J·m^{-2}; $SP = 6.3 \times 10^{-6}$; and $V_M = 4 \times 10^{-5}$ m^3·mol^{-1} [62]

higher pH values imply higher driving forces and, consequently, lower critical sizes. However, it is noted as an inflection dependent on the initial concentration, showing small changes in r_{crit} after this point. It is important to note that the critical size was observed below the nanometric range only in very diluted conditions. In more concentrated solutions or moderate pH conditions, the critical nucleus tends to higher values to become stable.

A similar analysis can be done to a poorly soluble amphoteric base. However, since the solubility will be strongly dependent on the pH, it is expected that the curve shows a maximum driving force, where it will coincide with the minimal r_{crit}. This was confirmed by the authors to $Pb(OH)_2$ [36].

Those examples show the importance of synthesis control to obtain desired sizes. The ideal nucleation condition, in order to obtain minimal particle sizes, is the precipitation in pH of maximum driving force, i.e., pH >12 for $Ca(OH)_2$ and diluted solutions.

An implication of those equations is seen in the way of precipitation reactions, like in sol-gel synthesis. The sol-gel process is defined as the production of nanoparticles by the hydrolysis of metal alkoxides, metal halides, and other inorganic salts. Equation 3.25 shows a typical hydrolytic reaction of an alkoxide compound [7]:

$$M - OR + H_2O \rightarrow M - OH \downarrow + ROH, \tag{3.25}$$

where M represents Si, Ti, Zr, Al, and other metals, R is a ligand such as an alkyl group or halide, and ROH is an alcohol. As shown above, the first step in the process is the precipitation of a hydroxide that can be understood here as a poorly soluble base. In the subsequent steps, the process will be controlled by polycondensation reactions to form the oxide. Since in the hydrolysis process the pH may change during the precipitation, depending on the experimental conditions adopted, the r_{crit} value may vary considerably during the process. In fact, several works show polydispersivity in the size distribution of nanoparticles synthesized by precipitation, despite in several cases near monodisperse distributions may be observed [37–41]. Also, an important factor is the variation of the ionic concentration, since when the particle is precipitated, the concentration in solution decays.

These two factors are implicitly evolved in the scheme shown in Fig. 3.5; in a solution with any concentration below the solubility limit c_s, the nucleation cannot occur (I). Over this point, the system becomes metastable, and some nuclei can be formed. At this point, the driving force for the nucleation is low, and the precipitation--redissolution will compete. Over c_{ss} (II), the system is supersaturated, and the nucleation will be rapid, relieving the concentration in solution. However, between II and III, several concentrations will be attained, and polydispersed nuclei will be formed. Reducing the driving force, the nucleation will stop, but the diffusion of ions or molecules onto the formed nuclei will continue, leading to the growth [42–45].

3.3 The Nucleation Rate

The dynamics of the nucleation process imply polydisperse particle distribution due to the factors discussed above, but also the nucleation rate is an important factor to define polydispersivity [46]. A general expression to define the equilibrium rate of nucleation per volume I_v may be proposed as follows [19, 20, 42]:

$$I_v = v \cdot n_s \cdot n_{emb}, \tag{3.26}$$

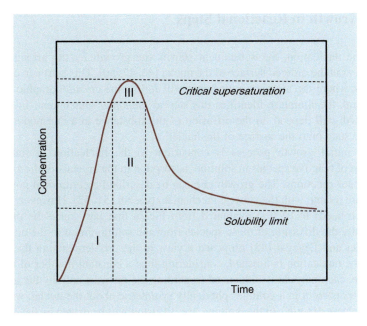

Fig. 3.5 General scheme of the precipitation of nuclei along the time (Adapted from Ref. [43])

where n_{emb} is the number of embryos (with a size smaller than the nuclei with critical size), n_s is the number of *monomers* in the vicinity of a single embryo, and v is their collision frequency with embryo surface. The term *monomer* is related to the coordinated structure or minimal crystalline form. Following the idea, in example, the *monomer* will be a TiO_6 octahedron for titanium oxide compounds or a SiO_4 tetrahedron for silica structures, or even single ions or molecules, depending of the system. In the reactional medium, the Brownian motion of the embryos may be explained by a Maxwell-Boltzmann statistics, i.e.:

$$n_{emb}/n_0 = \exp(-\Delta G_{crit}/kT), \tag{3.27}$$

where n_0 is the total number of ions or molecules and ΔG_{crit} is given by Eq. 3.6. The collision frequency in solution is often described also in the form $v = v_0 \cdot \exp(-\Delta G_m/kT)$, where v_0 is the molecular jump frequency and ΔG_m is the activation energy for transport in the interface. Substituting the terms in Eq. 3.26, one has:

$$I_v = v_0 \cdot n_s \cdot n_0 \cdot \exp(-\Delta G_{crit}/kT) \cdot \exp(-\Delta G_m/kT). \tag{3.28}$$

This expression shows the exponential dependence of the temperature in nucleation rate, and it is valid for nucleation in melts or reactional media. It is important to note that the expression did not evaluate the effect of the redissolution of supercritical nuclei; however, further corrections did not change this general profile [19, 20, 35].

3.4   Growth in Reactional Steps

After the nucleation, the subsequent step is the growth by the arrangement of *monomers* in the ordered surface of the nuclei [43, 47–50]. This is true in crystalline systems, where the atomic arrangement will follow the crystallographic ordering. In general, the literature identifies this step as *monomer* attachment. In any case, the growth will depend on the diffusion of the substance in a thermodynamically unstable state onto the surface of the nuclei.

This initial growth process is concurrent to the nucleation in some ways: transport of reactive species in solution, adsorption in the crystal-solution interface, or interface reactions. The growth rate can be described by empirical power laws, which are related with the slowest mechanism present [51, 52]. However, a general approach needs to be first evaluated from Fick's first law, since the problem is essentially the diffusion of those species from the solution/melt to the nucleus.

LaMer and Dinegar [43] proposed a view to the problem stating the situation where the nucleation is finished, considering then a constant number of nuclei. In this case, each particle is influenced (regarding the growth) only by the amount of *monomers* present in a volume spherically symmetric about the nuclei, with radius h and volume $4/3 \cdot \pi \cdot h^3$. In this volume, the diffusion will be driven by the difference between the concentration of *monomers* in solution and the solubility concentration of the solid phase, i.e., $(C_{ss} - C_s) = C_0$. Since the solubility is a temperature function, the proper equation can be estimated in principle considering isothermal conditions. Then, following Fick's law, one can describe the flux J through a spherical volume:

$$J(t) = -4\pi r^2 D \, dC/dr, \tag{3.29}$$

where D corresponds to the diffusion coefficient, also dependent on temperature. The following boundary conditions are assumed: (i) at the particle surface, the concentration has the saturation value C_s; (ii) in the instant $t = 0$, the concentration at any point is the supersaturation value C_{ss}; (iii) in $r = h$, the variation of the concentration with the time is dependent only to the flux J, i.e., $(\delta C/\delta t)_{r=h} = 3 \cdot J(t)/4\pi h^3$. Equation 3.29 may be integrated to obtain:

$$C(r,t) = J(t)/(4\pi \cdot r \cdot D) + f(t), \tag{3.30}$$

since $t = 0$, $J(0) = 0$ and $C(r,0) = f(0) = C_{ss}$, from condition (ii). Deriving the Eq. 3.30 in respect to t allows us to write:

$$(\delta C/\delta t)_{r=h} = 1/(4\pi \cdot h \cdot D) \cdot dJ/dt + df/dt. \tag{3.31}$$

Comparing the equation above with condition (iii), and integrating the term df/dt, one has:

$$f(t) - f(0) = 1/(4\pi \cdot h \cdot D) \cdot (J(t) - J(0)) + 3/(4\pi \cdot h^3) \cdot \int_0^t J(t) dt, \tag{3.32}$$

since $J(0) = 0$, and from condition (ii) $f(0) = C_{ss}$, the equation may be simplified to a solution for $f(t)$ that can be applied in Eq. 3.30:

$$C(r,t) = C_{ss} + J(t)/(4\pi \cdot D) \cdot (1/r - 1/h) + 3/(4\pi \cdot h^3) \cdot \int_0^t J(t)dt. \quad (3.33)$$

Assuming $h \gg r$, the equation may be simplified to:

$$C(r,t) = C_{ss} + J(t)/(4\pi \cdot D \cdot r) + 3/(4\pi \cdot h^3) \cdot \int_0^t J(t)dt. \quad (3.34)$$

The total flux over the particle is equal to the increase of the particle volume, i.e., $J(t)dt = \rho d(4/3\pi \cdot r^3) = 4\pi\rho r^2 dr$, where r is the particle radius and ρ is the density of the particle. Then, Eq. 3.34 can be rewritten as:

$$C(r,t) = C_{ss} + \rho \cdot r/D \cdot dr/dt + 3\rho/h^3 \int_0^r r^2 dr. \quad (3.35)$$

The integration limits in the last term were substituted since in any time the radius of the particle is r, and in $t = 0$, the particle has the critical size; however, for the sake of simplicity, it is adopted arbitrarily $r_{initial} = 0$. Although the approximation is not true, the result will focus only the growth process, and the analysis will be satisfactory. After integrating and rearranging the expression, one has:

$$dr/dt = D/\rho \ (C_{ss} - C_s/r - \rho r^2/h^3). \quad (3.36)$$

The equation may be analytically solved if the term $(C_{ss}-C_s) = C_0$ is considered constant with the time. This is a plausible condition in very concentrated suspensions or in conditions where the driving force for nucleation is high, like in the hydrolysis in elevated pHs. The solution for this special case is:

$$t = B \cdot 1/(6A) \cdot \log[(A^2 + Ar + r^2)/(A - r)^2] - 1/(A\sqrt{(3)})$$
$$\cdot \tan^{-1}[(2r+A)/(A\sqrt{(3)})], \quad (3.37)$$

where $A = C_0 \cdot h^3/\rho$ and $B = h^3/D$. All the terms in the equation are possible to evaluate: C_0 and D are measured experimentally, and h can be estimated by measuring the total number of particles in solution. A schematic plot of Eq. 3.37 is seen in Fig. 3.6. The observed shape is approximately a logarithmic curve, showing a rapid growth in initial stages, tending to a stable value.

The previous analysis shows that control of the subsequent growth of as-formed nuclei is difficult, since the evolution of the sizes is very fast. This result is strongly

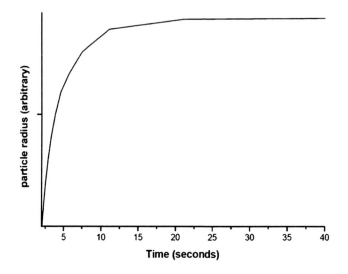

Fig. 3.6 Plot of Eq. 3.37. All the constants were assumed as unit

influenced by the supersaturation value, and this is maybe the most important variable to control in the synthesis. As discussed above, one can see the value C_0 as an indicative of the driving force for nucleation, since the value C_{ss} is a function of temperature and pH. Then, severe changes in pH can improve the value even in very dilute systems, leading to the fast growth but limited to the total solute content. In this case, the curve will settle rapidly to a stable value, since there is no disposable material in solution to the growth.

The general discussion shows that, to obtain nanoparticles from solutions, it is usually necessary to stop the growth mechanisms or at least to control them to prevent uncontrollable growth and, hence, undesirable particle sizes. All the reactional parameters can be controlled by the proper selection of reactant relations [36, 53]. As an example, in precipitation by hydrolysis, a large excess of water in relation to the metal source reactant leads to nanoparticles due to the fact that all the *monomers* present in solution are captured in primary nucleus [7, 13, 54–60]. This strategy was used to produce Sb:SnO$_2$ nanoparticles near monodisperse by fast hydrolysis, as showed in high-resolution transmission microscopy (HRTEM) image in Fig. 3.7.

Once again, control of growth after nucleation is necessary to obtain desirable nanoparticles. Wu et al. [61] used an interesting approach to control the particle growth of hydrous metal oxide gels. They showed that the growth could be inhibited by replacing the surface hydroxyl group, before the crystallization step, with a functional group that does not condense and that can produce small secondary-phase particles which restrict boundary mobility during the synthesis and even at high temperatures. These authors reported that fully crystalline SnO$_2$, TiO$_2$, and ZrO$_2$ nanocrystals (ranging in size from 1.5 to 5 nm) can be obtained after heat treating the precipitate gel at 500°C, by replacing the hydroxyl group with the methyl siloxyl group before firing.

Fig. 3.7 HRTEM image of Sb:SnO$_2$ nanoparticles synthesized by fast hydrolysis of solution of metal halides

3.5 Heterogeneous Nucleation

The set of equations can be corrected to consider the reduction of surface energy by contact with other surfaces (which represents a *heterogeneous nucleation* process) by substituting γ to γ_{eff}, an effective surface energy obtained by contact with another surface [62, 63]. Applying the geometric relations of the contact with a cluster and a surface, one can consider the decrease of activation energy to nucleation (as given in Eq. 3.6), i.e.:

$$\Delta G_{\text{crit,het}} = \Delta G_{\text{crit}}(2 + \cos\theta)(1 - \cos\theta)^2/4, \qquad (3.38)$$

where θ is the contact angle with the surfaces (heterogeneity) and the medium. The heterogeneous nucleation is preferable only if the geometric relation is less than 1.

In general, most of the nucleation process is heterogeneous, since simple interferences (like contaminants) can be embryos. Practical application of Eq. 3.38 is difficult, since in many cases the contact angle is undetermined. However, large surfaces also act as heterogeneities, and in this case, the event is easily identified. The wettability of the surface in the reactional media is a direct evidence of the heterogeneous nucleation: if the solution can wet the surface, the angle θ will be consequently lower than $\pi/2$, and the process will be preferable. In this sense, even the surfaces of containers where the reaction takes place may act as the heterogeneity.

Following this idea, it is possible to tailor structures from surfaces, like thin films, using a proper substrate immersed in the reactional media [64–74], as exemplified in Fig. 3.8 for the synthesis of SnO$_2$ films by heterogeneous nucleation over Sb:SnO$_2$ polycrystalline substrate. The substrate decreases the activation energy in the crystallization process, promoting heterogeneous nucleation at the

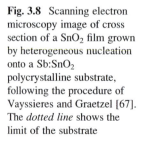

Fig. 3.8 Scanning electron microscopy image of cross section of a SnO_2 film grown by heterogeneous nucleation onto a $Sb:SnO_2$ polycrystalline substrate, following the procedure of Vayssieres and Graetzel [67]. The *dotted line* shows the limit of the substrate

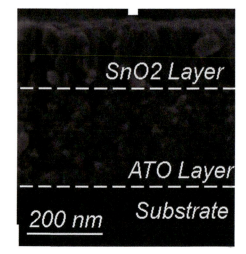

surface of the solid. The kinetics of nucleation and growth may be controlled by the temperature and the hydrolysis rate, whereby the nucleation and growth processes can be separated.

3.6   Deviations of the Models: The Determination of γ

The models discussed previously were extensively used to explain nucleation and growth of microscopic particles, but they failed in many cases when applied to nanoscopic objects. One of the major error factors is the shape assumption – sphere – that is not correct for very small objects. Experimental [75–81] and theoretical works [9, 82–88] shows the spontaneous formation of anisotropic shapes of nanoparticles, and in some cases, with faceted shapes, as shown in Fig. 3.9 for Fe_3O_4 synthesized in organic solvent. Solid surfaces of different crystallographic orientations have different surface energies and different affinities for absorbed ions and molecules [89]. Then, shape is an important variable in the nucleation of nanocrystals, since the correct surface energy depends on the exposed crystallographic planes [90].

Barnard and Zapol [9] proposed a general model for the phase stability of any nanoparticle based on the Gibbs free energy of an arbitrary particle. According to the authors, the correct treatment of the free energy must include contributions from the edges and corners rather than only from the bulk and surface. As an example, a Si nanocrystals (with cubic structure) with 200 atoms will have 9% of the atoms in edges; with 10^3 atoms, 4% will be in the edges; and with 10^5, only 0.3% will remain. These estimates highlight the importance of the small terms, in some cases. For a given x nanoparticle, the free energy can be expressed as a sum of individual contributions, i.e.:

$$G_x^{\,0} = G_x^{\,\text{bulk}} + G_x^{\,\text{surface}} + G_x^{\,\text{edge}} + G_x^{\,\text{corners}}. \tag{3.39}$$

Fig. 3.9 HRTEM image of faceted Fe_3O_4 nanoparticles synthesized in organic solvent

The first term is defined as the standard free energy of formation, $G_x^{\text{bulk}} = \Delta G_x^0(T)$, which is dependent on the temperature. The second term is expressed in terms of surface energy γ_i for each i plane on the surface and molar surface area A. Using the relations of density ρ, molar mass M, and surface to volume ratio q, one has:

$$G_x^{\text{surface}} = \gamma(T)A = (M/\rho) \cdot q\Sigma_i f_i \gamma_i(T), \tag{3.40}$$

where f_i is a weight factor of the facets i in the crystal ($\Sigma f_i = 1$). In the above formulation, the expression takes into account the crystallographic alignment of the properties and, indirectly, the shape. The edge and corner energies can be described by similar expressions:

$$G_x^{\text{edge}} = \xi(T)L = (M/\rho)p\Sigma_j g_j\ \xi_j(T), \tag{3.41}$$

$$G_x^{\text{corner}} = \tau(T)W = (M/\rho)w\Sigma_k h_k\tau_k(T), \tag{3.42}$$

where $\xi(T)$, $\tau(T)$ are the edge and corner free energies, L is the total length of edges and W is the total number of corners, p, w the edge and the corner to volume ratios, and g_j, h_k weight factors. Substituting and rearranging the terms, Eq. 3.39 becomes:

$$G_x^0 = \Delta G_x^0(T) + (M/\rho)\left[q\Sigma_i f_i\gamma_i(T) + p\Sigma_j g_j\xi_j(T) + w\Sigma_k h_k\tau_k(T)\right]. \tag{3.43}$$

However, the effective pressure must be taken into account [91]. The volume dilation e_d is given as:

$$\Delta V/V = e_d = P_{eff}\beta_V, \tag{3.44}$$

where β_V is the material's compressibility. P_{eff} can be estimated by Eq. 3.3 above. Although it is known that $\sigma = \gamma + A(\partial\gamma/\partial A)$, when the dependence of γ on A is small, the approximation $\gamma = \sigma$ is acceptable. The anisotropy should be included in the determination of γ, as done in Eq. 3.40. Using these approximations, Eq. 3.43 becomes:

$$G_x{}^0 = \Delta G_x{}^0 + (M/\rho)(1 - 2\beta_V\sigma/R)[q\Sigma f_i\gamma_i + p\Sigma g_j\xi_j + w\Sigma h_k\tau_k]. \tag{3.45}$$

Although the Laplace-Young equation (Eq. 3.3) is only applicable to spherical particles, the approach was tested successfully in simulations for Si, Ge, nano-diamonds, and TiO_2 polymorphs. The authors predicted faceted shapes (e.g., cubes or tetrakaidecahedrons) as preferred shapes in very small sizes for most cases, which is contradictory to common sense (i.e., spheres) [9, 83, 85].

This strong dependence on surface, corner, and edge energies can modify significantly the analysis of the nucleation and growth processes. Two approaches may be evaluated: consider an "average" surface energy, given by the distribution of individual surface energies, or assume each surface as an independent site during the growth. This second approach was observed as valid in experimental works regarding synthesis of nanorods or nanowires [92–96]. In this case, the growth process needs be understood as independent in each crystallographic direction. However, those assumptions make the interpretation of the nucleation and growth a very complex problem.

3.7 Crystal Growth

Despite the growth can occur in reactional steps, this topic will be focused in the events occurring after the reactional equilibrium, i.e., growth phenomena occurring after the reactants consume but dependent on diffusional parameters and on the particles' relative mobility. Some theories are reviewed here, focusing the Ostwald ripening and oriented attachment theories.

3.8 Ostwald Ripening

Usually, nanoparticle growth in equilibrium condition is associated with *coarsening*, which is also known as *Ostwald ripening*. This mechanism can be described as a diffusion or reaction rate–limited growth of nanoparticles at the expense of smaller

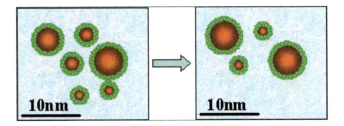

Fig. 3.10 Scheme of the Ostwald ripening growth: bigger particles grown at the expense of smaller ones

ones [97, 98], as schematically shown in Fig. 3.10. A kinetic model for the *Ostwald ripening* mechanism was rigorously developed by Lifshitz, Slyozov [99], and Wagner [100] and is also known as the LSW model. The LSW model predicts that the mean particle radius should evolve as a function of time according to the following equation:

$$R_{p,m}^{n} - R_{p,m,o}^{n} \propto t, \tag{3.46}$$

where n is dependent of the limiting step to the growth. This result is obtained by combining the Gibbs-Thompson equation – which describes the dependence of the particle solubility as a function of its size – and Fick's first law. The power law coefficient $n = 3$ is obtained by considering dilute conditions, where diffusion of ions in solution is the limiting step. In concentrated conditions (e.g., solids, or nucleation in melts), the coefficient may be equivalent to 2 or 4, depending on the limiting step involved in the interfacial reactions (i.e., dissolution or reprecipitation) [101, 102].

A didactic development of the equation is given by some simplified assumptions, as follows. Initially, Eq. 3.17 can be rearranged assuming the activity a is equal to the solubility of the formed particle S_p, and the saturation activity a_0 as the bulk solubility $S_{b,0}$ [103]:

$$S_p = S_{b,0} . \exp[2V_m \gamma / (RTr)], \tag{3.47}$$

This relation, widely known as the *Ostwald-Freundlich* equation, describes the dependence of the solubility of a formed particle on its size. This dependence is particularly important in very small particles, since dissolution and reprecipitation phenomena can easily occur [104]. Usually, the argument in exponential terms is too small, and the equation can be simplified to:

$$S_p \approx S_{b,0} \cdot (1 + 2V_m \gamma / RT) \cdot (1/r). \tag{3.48}$$

If particles dissolve and grow readily without being limited by the rate of interfacial reactions, the growth rate of the particles is likely to be limited by diffusion through a surrounding medium and can be described by Fick's first law.

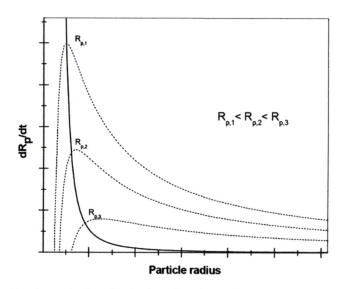

Fig. 3.11 Growth rate in Ostwald ripening of particles with arbitrary radius, according to Eq. 3.50. The *straight line* refers to the maximum growth rate, when $R_p = 2R_p$

This supposition is clearly valid in colloids and in reactional media. A convenient form of Fick's first law for a particle in a diffusion field is given as (in polar coordinates) [105]:

$$4\pi r^2 dr/dt = D4\pi x^2 dc/dx, \tag{3.49}$$

where D is the diffusivity and dc/dx is the gradient in concentration at distance x. Considering that, in a distance $x \gg r$, the solubility is the same as that of the average particle size ($R_{p,m}$), Eq. 3.49 can be rewritten for evaluation at $x = R_p$, after integration of the right-hand side:

$$\begin{aligned} dR_p/dt &= -D/R_p[c(R_p) - c(R_{p,m})] \\ &= -(c_0 D/R_p)(2V_m\gamma/RT)(1/R_p - 1/R_{p,m}). \end{aligned} \tag{3.50}$$

The concentration values are substituted by using Eq. 3.47, assuming $c = S_p$ and $c_0 = S_{b,0}$. Figure 3.11 shows a schematic distribution of growth rates for some arbitrary values of particle radii. Clearly, the maximum growth rate will occur in a defined range of particles. Taking the second derivative $d^2R_p/dt^2 = 0$, it is observed that, when $R_p = 2R_p$, one has the condition of maximum growth rate, as plotted in Fig. 3.11 by the straight line. Assuming that growth (in a closed system) is governed by the fastest growing particles, one can write $dR_p = dR_{p,m}$. Substituting these values in Eq. 3.50 and integrating them, a particular form of Ostwald ripening is obtained [20, 62, 106]:

$$R_{p,m}^3 - R_{p,m,0}^3 = (3c_0 DV_m\gamma/4RT) \cdot t. \tag{3.51}$$

This growth mechanism provides a good description of the growth behavior of a wide range of nanoparticles [97, 98, 101, 107–114].

Going to equation LSW, the exponent n is related to the boundary conditions assumed in the growth [100, 102]: for $n = 2$, it is inferred that crystal growth is controlled by ion diffusion throughout the particle to its vicinity (the solution of a matrix, in solid compounds); for $n = 3$, the growth is controlled by the volume diffusion of ions in the vicinity, and when $n = 4$, it is deduced that growth is controlled by dissolution kinetics at the particle-vicinity interface.

References

1. Adair, J.H., Suvaci, E.: Morphological control of particles. Curr. Opin. Colloid Interface Sci. **5**, 160–167 (2000)
2. Cushing, B.L., Kolesnichenko, V.L., O'Connor, C.J.: Recent advances in the liquid-phase syntheses of inorganic nanoparticles. Chem. Rev. **104**, 3893–3946 (2004)
3. Toschev, S.: Homogeneous nucleation. In: Hartman, P. (ed.) Crystal Growth – An Introduction Homogeneous nucleation. Elsevier, New York, pp. 1–49. (1973)
4. von Weimarn, P.P.: The precipitation laws. Chem. Rev. **2**, 217–242 (1925)
5. Ludwig, F.-P., Schmelzer, J.: On von Weimarn's law in nucleation theory. J. Colloid Interface Sci. **181**, 503–510 (1996)
6. Zhang, H.Z., Banfield, J.F.: Phase transformation of nanocrystalline anatase-to-rutile via combined interface and surface nucleation. J. Mater. Res. **15**, 437–448 (2000)
7. Brinker, C.J., Scherrer, G.W.: Sol-Gel Science. Academic Press, Boston (1990)
8. Zhang, Y., Li, Y., Kim, W., Wang, D., Dai, H.: Imaging as-grown single-walled carbon nanotubes originated from isolated catalytic nanoparticles. Appl. Phys. A: Mater. Sci. Proc. **74**, 325–328 (2002)
9. Barnard, A.S., Zapol, P.: A model for the phase stability of arbitrary nanoparticles as a function of size and shape. J. Chem. Phys. **121**, 4276–4283 (2004)
10. Alivisatos, A.P.: Perspectives on the physical chemistry of semiconductor nanocrystals. J. Phys. Chem. **100**, 13226–13239 (1996)
11. Tolbert, S.H., Alivisatos, A.P.: Size dependence of a first-order solid-solid phase-transition - the wurtzite to rock-salt transformation in CdSe nanocrystals Science. Science **265**, 373–376 (1994)
12. Tolbert, S.H., Alivisatos, A.P.: The wurtzite to rock-salt structural transformation in CdSe nanocrystals under high-pressure. J. Chem. Phys. **102**, 4642–4656 (1995)
13. Livage, J., Henry, M., Sanchez, C.: Sol-gel chemistry of transition-metal oxides. Prog. Solid State Chem. **18**, 259–341 (1988)
14. Matijevic, E.: Uniform inorganic colloid dispersions. Achievements and challenges. Langmuir **10**, 8–16 (1994)
15. Matijevic, E.: Preparation and properties of uniform size colloids. Chem. Mater. **5**, 412–426 (1993)
16. Matijevic, E.: Monodispersed colloids: art and science. Langmuir **2**, 12–20 (1986)
17. Matijevic, E.: Monodispersed metal (hydrous) oxides – a fascinating field of colloid science. Acc. Chem. Res. **14**, 22–29 (1981)
18. Paul, A.: Chemistry of Glasses. Chapman and Hall, London/New York (1982)
19. Kingery, W.D., Bowen, H.K., Uhlmann, D.R.: Introduction to Ceramics, 2nd edn. John Wiley and Sons, New York (1976)
20. Chiang, Y.M., Birnie, D.P., Kingery, W.D.: Physical Ceramics. John Wiley and Sons, New York (1997)
21. Navrotsky, A., Kleppa, O.J.: Enthalpy of anatase-rutile transformation. J. Am. Ceram. Soc. **50**, 626 (1967)

22. Ohtaka, O., Yamanaka, T., Kume, S., Ito, E., Navrotsky, A.: Stability of monoclinic and orthorhombic zirconia – studies by high-pressure phase-equilibria and calorimetry. J. Am. Ceram. Soc. **74**, 505–509 (1991)
23. McHale, J.M., Auroux, A., Perrotta, A.J., Navrotsky, A.: Surface energies and thermodynamic phase stability in nanocrystalline aluminas. Science **277**, 788–791 (1997)
24. Ranade, M.R., Navrotsky, A., Zhang, H.Z., Banfield, J.F., Elder, S.H., Zaban, A., Borse, P.H., Kulkarni, S.K., Doran, G.S., Whitfield, H.J.: Energetics of nanocrystalline TiO_2. Proc. Natl. Acad. Sci. U.S.A. **99**, 6476–6481 (2002)
25. Ushakov, S.V., Navrotsky, A.: Direct measurements of water adsorption enthalpy on hafnia and zirconia. Appl. Phys. Lett. **87**, 164103 (2005)
26. Pitcher, M.W., Ushakov, S.V., Navrotsky, A., Woodfield, B.F., Li, G., Boerio-Goates, J., Tissue, B.M.: Energy crossovers in nanocrystalline zirconia. J. Am. Ceram. Soc. **88**, 160–167 (2005)
27. Rimer, J.D., Trofymluk, O., Navrotsky, A., Lobo, R.F., Vlachos, D.G.: Kinetic and thermodynamic studies of silica nanoparticle dissolution. Chem. Mater. **19**, 4189–4197 (2007)
28. Majzlan, J., Mazeina, L., Navrotsky, A.: Enthalpy of water adsorption and surface enthalpy of lepidocrocite ([gamma]-FeOOH). Geochim. Cosmochim. Acta **71**, 615–623 (2007)
29. Navrotsky, A.: Energetic clues to pathways to biomineralization: Precursors, clusters, and nanoparticles. Proc. Natl. Acad. Sci. U.S.A. **101**, 12096–12101 (2004)
30. Interrante, L.V., Hampden-Smith, M.J.: Chemistry of advanced materials. Wiley-VCH, New York (1998)
31. Mutaftschiev, B.: Nucleation. In: Hurle, D.T.J. (ed.) Handbook of Crystal Growth, p. 187. Elsevier Science, Amsterdam (1993)
32. Navarro, J.M.F.: El Vidrio, 2nd edn. C.S.I.C., Madrid (1991)
33. Kingery, W.D.: Densification during sintering in the presence of a liquid phase.1. Theory. J. Appl. Phys. **30**, 301–306 (1959)
34. Boudart, M., Djega-Mariadasson, G.: Kinetics of heterogeneous reaction. Princeton University Press, Princeton (1981)
35. Fokin, V.M., Zanotto, E.D., Yuritsyn, N.S., Schmelzer, J.W.: Homogeneous crystal nucleation in silicate glasses: A 40 years perspective. J. Non. Cryst. Solids **352**, 2681–2714 (2006)
36. Kukushkin, S.A., Nemna, S.V.: The effect of pH on nucleation kinetics in solutions. Dokl. Phys. Chem. **377**, 792–796 (2001)
37. Barringer, E.A., Bowen, H.K.: High-purity, monodisperse TiO_2 powders by hydrolysis of titanium tetraethoxide. 1. Synthesis and physical-properties. Langmuir **1**, 414–420 (1985)
38. Jean, J.H., Ring, T.A.: Nucleation and growth of monosized TiO_2 powders from alcohol solution. Langmuir **2**, 251–255 (1986)
39. Bahnemann, D.W., Kormann, C., Hoffmann, M.R.: Preparation and characterization of quantum size zinc-oxide – a detailed spectroscopic study. J. Phys. Chem. **91**, 3789–3798 (1987)
40. Kormann, C., Bahnemann, D.W., Hoffmann, M.R.: Preparation and characterization of quantum-size titanium dioxide. J. Phys. Chem. **92**, 5196–5201 (1988)
41. van Blaaderen, A., Vrij, A.: Synthesis and characterization of monodisperse colloidal organosilica spheres. J. Colloid Interface Sci. **156**, 1–18 (1993)
42. Lamer, V.K.: Nucleation in phase transitions. Ind. Eng. Chem. **44**, 1270–1277 (1952)
43. Lamer, V.K., Dinegar, R.H.: Theory, production and mechanism of formation of monodispersed hydrosols. J. Am. Chem. Soc. **72**, 4847–4854 (1950)
44. Pound, G.M., Lamer, V.K.: Kinetics of crystalline nucleus formation in supercooled liquid tin. J. Am. Chem. Soc. **74**, 2323–2332 (1952)
45. Reiss, H., Lamer, V.K.: Diffusional boundary value problems involving moving boundaries, connected with the growth of colloidal particles. J. Chem. Phys. **18**, 1–12 (1950)
46. Alekseechkin, N.: Multidimensional kinetic theory of first-order phase transitions. Phys. Solid State **48**, 1775–1785 (2006)
47. Weiss, G.H.: Overview of theoretical models for reaction rates. J. Stat. Phys. **42**, 3–36 (1986)

48. Bradley, J.S.: The chemistry of transition metal colloids. In: Schmid, G. (ed.) Clusters and Colloids: From Theory to Applications, pp. 459–544. VCH, Weinheim (1994)

49. Givargizov, E.I.: Oriented Crystallization on Amorphous Substrates. Plenum Press, New York (1991)

50. Jensen, P.: Growth of nanostructures by cluster deposition: Experiments and simple models. Rev. Mod. Phys. **71**, 1695–1735 (1999)

51. Aoun, M., Plasari, E., David, R., Villermaux, J.: A simultaneous determination of nucleation and growth rates from batch spontaneous precipitation. Chem. Eng. Sci. **54**, 1161–1180 (1999)

52. Zauner, R., Jones, A.G.: Determination of nucleation, growth, agglomeration and disruption kinetics from experimental precipitation data: The calcium oxalate system. Chem. Eng. Sci. **55**, 4219–4232 (2000)

53. Leite, E.R.: Nanocrystals assembled from bottom-up. In: Nalwa, H.S. (ed.) Encyclopedia of Nanoscience and Nanotechnology, pp. 537–550. American Scientific Publishers, Stevenson Ranch (2004)

54. Liu, C., Zou, B.S., Rondinone, A.J., Zhang, Z.J.: Sol-gel synthesis of free-standing ferroelectric lead zirconate titanate nanoparticles. J. Am. Chem. Soc. **123**, 4344–4345 (2001)

55. Lee, J., Isobe, T., Senna, M.: Preparation of ultrafine Fe_3O_4 particles by precipitation in the presence of PVA at high pH. J. Colloid Interface Sci. **177**, 490–494 (1996)

56. Shevchenko, E.V., Talapin, D.V., Schnablegger, H., Kornowski, A., Festin, O., Svedlindh, P., Haase, M., Weller, H.: Study of nucleation and growth in the organometallic synthesis of magnetic alloy nanocrystals: The role of nucleation rate in size control of $CoPt_3$ nanocrystals. J. Am. Chem. Soc. **125**, 9090–9101 (2003)

57. Oskam, G., Poot, F.D.P.: Synthesis of ZnO and TiO_2 nanoparticles. J. Sol-Gel Sci. Technol. **37**, 157–160 (2006)

58. Segal, D.: Chemical synthesis of ceramic materials. J. Mater. Chem. **7**, 1297–1305 (1997)

59. Wang, C.C., Ying, J.Y.: Sol-gel synthesis and hydrothermal processing of anatase and rutile titania nanocrystals. Chem. Mater. **11**, 3113–3120 (1999)

60. Ying, J.Y.: Preface to the special issue: Sol-gel derived materials. Chem. Mater. **9**, 2247–2248 (1997)

61. Wu, N.L., Wang, S.Y., Rusakova, I.A.: Inhibition of crystallite growth in the sol-gel synthesis of nanocrystalline metal oxides. Science **285**, 1375–1377 (1999)

62. Chakraverty, B.K.: Heterogeneous nucleation and condensation on substrates. In: Hartman, P. (ed.) Crystal Growth – An Introduction, pp. 50–104. Elsevier, New York (1973)

63. Cacciuto, A., Auer, S., Frenkel, D.: Onset of heterofeneous crystal nucleation in colloidal suspensions. Nature **428**, 404–406 (2004)

64. Vayssieres, L.: Growth of arrayed nanorods and nanowires of ZnO from aqueous solutions. Adv. Mater. **15**, 464–466 (2003)

65. Vayssieres, L., Beermann, N., Lindquist, S.E., Hagfeldt, A.: Controlled aqueous chemical growth of oriented three-dimensional crystalline nanorod arrays: Application to iron(III) oxides. Chem. Mater. **13**, 233–235 (2001)

66. Vayssieres, L., Chaneac, C., Tronc, E., Jolivet, J.P.: Size tailoring of magnetite particles formed by aqueous precipitation: An example of thermodynamic stability of nanometric oxide particles. J. Colloid Interface Sci. **205**, 205–212 (1998)

67. Vayssieres, L., Graetzel, M.: Highly ordered SnO_2 nanorod arrays from controlled aqueous growth. Angew. Chem. Int. Ed. **43**, 3666–3670 (2004)

68. Vayssieres, L., Hagfeldt, A., Lindquist, S.E.: Purpose-built metal oxide nanomaterials. The emergence of a new generation of smart materials. Pure Appl. Chem. **72**, 47–52 (2000)

69. Vayssieres, L., Keis, K., Hagfeldt, A., Lindquist, S.E.: Three-dimensional array of highly oriented crystalline ZnO microtubes. Chem. Mater. **13**, 4395 (2001)

70. Vayssieres, L., Keis, K., Lindquist, S.E., Hagfeldt, A.: Purpose-built anisotropic metal oxide material: 3D highly oriented microrod array of ZnO. J. Phys. Chem. B **105**, 3350–3352 (2001)

71. Vayssieres, L., Manthiram, A.: 2-D mesoparticulate arrays of alpha-Cr_2O_3. J. Phys. Chem. B **107**, 2623–2625 (2003)
72. Vayssieres, L., Rabenberg, L., Manthiram, A.: Aqueous chemical route to ferromagnetic 3-d arrays of iron nanorods. Nano Lett. **2**, 1393–1395 (2002)
73. Vayssieres, L., Sathe, C., Butorin, S.M., Shuh, D.K., Nordgren, J., Guo, J.H.: One-dimensional quantum-confinement effect in alpha-Fe_2O_3 ultrafine nanorod arrays. Adv. Mater. **17**, 2320 (2005)
74. Zhang, H., Wang, D.Y., Yang, B., Mohwald, H.: Manipulation of aqueous growth of CdTe nanocrystals to fabricate colloidally stable one-dimensional nanostructures. J. Am. Chem. Soc. **128**, 10171–10180 (2006)
75. Cheng, B., Russel, J.M., Shi, W., Zhang, L., Samulski, E.T.: Large-scale, solution-phase growth of single-crystalline SnO_2 nanorods. J. Am. Chem. Soc. **126**, 5972–5973 (2004)
76. Cozzoli, P., Kornowski, A., Weller, H.: Low-temperature synthesis of soluble and processable organic-capped anatase TiO_2 nanorods. J. Am. Chem. Soc. **125**, 14539–14548 (2003)
77. Joo, J., Kwon, S.G., Yu, T., Cho, M., Lee, J., Yoon, J., Hyeon, T.: Large-scale synthesis of TiO_2 nanorods via nonhydrolytic sol-gel ester elimination reaction and their application to photocatalytic inactivation of E. coli. J. Phys. Chem. B **109**, 15297–15302 (2005)
78. Liu, J.P., Huang, X.T., Sulieman, K.M., Sun, F.L., He, X.: Solution-based growth and optical properties of self-assembled monocrystalline ZnO ellipsoids. J. Phys. Chem. B **110**, 10612–10618 (2006)
79. Pacholski, C., Kornowski, A., Weller, H.: Self-assembly of ZnO: From nanodots, to nanorods. Angew. Chem. Int. Ed. **41**, 1188 (2002)
80. Park, S.J., Kim, S., Lee, S., Khim, Z.G., Char, K., Hyeon, T.: Intercalation of magnesium-urea complex into swelling clay. J. Am. Chem. Soc. **122**, 8581 (2004)
81. Talapin, D.V., Shevchenko, E.V., Murray, C.B., Kornowski, A., Forster, S., Weller, H.: CdSe and CdSe/CdS nanorod solids. J. Am. Chem. Soc. **126**, 12984–12988 (2004)
82. Barnard, A.: A thermodynamic model for the shape and stability of twinned nanostructures. J. Phys. Chem. B **110**, 24498–24504 (2006)
83. Barnard, A., Curtiss, L.: Prediction of TiO_2 nanoparticle phase and shape transitions controlled by surface chemistry. Nano Lett. **5**, 1261–1266 (2005)
84. Barnard, A., Saponjic, Z., Tiede, D., Rajh, T., Curtiss, L.: Multi-scale modeling of titanium dioxide: controlling shape with surface chemistry. Rev. Adv. Mater. Sci. **10**, 21–27 (2005)
85. Barnard, A.S., Yeredla, R.R., Xu, H.: Modelling the effect of particle shape on the phase stability of ZrO_2 nanoparticles. Nanotechnology **17**, 3039–3047 (2006)
86. Barnard, A.S., Zapol, P.: Predicting the energetics, phase stability, and morphology evolution of faceted and spherical anatase nanocrystals. J. Phys. Chem. B **108**, 18435–18440 (2004)
87. Barnard, A.S., Zapol, P.: Effects of particle morphology and surface hydrogenation on the phase stability of TiO_2. Phys. Rev. B: Condens. Matter Mater. Phys. **70**, 235403-1-13 (2004)
88. Barnard, A.S., Zapol, P., Curtiss, L.A.: Modeling the morphology and phase stability of TiO_2 nanocrystals in water. J. Chem. Theory Comput. **1**, 107–116 (2005)
89. Herring, C.: Some theorems on the free energies of crystal surfaces. Phys. Rev. **82**, 87–93 (1951)
90. Yacaman, M.J., Ascencio, J.A., Liu, H.B., Gardea-Torresdey, J.: Structure shape and stability of nanometric sized particles. J. Vac. Sci. Technol. B **19**, 1091–1103 (2001)
91. Zhang, H.Z., Banfield, J.F.: Kinetics of crystallization and crystal growth of nanocrystalline anatase in nanometer-sized amorphous titania. Chem. Mater. **14**, 4145–4154 (2002)
92. Tang, Z.Y., Kotov, N.A., Giersig, M.: Spontaneous organization of single CdTe nanoparticles into luminescent nanowires. Science **297**, 237–240 (2002)
93. Jun, Y.W., Casula, M.F., Sim, J.H., Kim, S.Y., Cheon, J., Alivisatos, A.P.: Surfactant-assisted elimination of a high energy facet as a means of controlling the shapes of TiO_2 nanocrystals. J. Am. Chem. Soc. **125**, 15981–15985 (2003)
94. Xi, G.C., Xiong, K., Zhao, Q.B., Zhang, R., Zhang, H.B., Qian, Y.T.: Nucleation-dissolution-recrystallization: A new growth mechanism for t-selenium nanotubes. Cryst. Growth Des. **6**, 577–582 (2006)

95. Mitra, S., Das, S., Mandal, K., Chaudhuri, S.: Synthesis of a alpha-Fe$_2$O$_3$ nanocrystal in its different morphological attributes: growth mechanism, optical and magnetic properties. Nanotechnology **18**, 275608 (2007)
96. Xu, X.X., Liu, F., Yu, K.H., Huang, W., Peng, B., Wei, W.: A kinetic model for nanocrystal morphology evolution. Chemphyschem **8**, 703–711 (2007)
97. Oskam, G., Hu, Z.S., Penn, R.L., Pesika, N., Searson, P.C.: Coarsening of metal oxide nanoparticles. Phys. Rev. E: Stat. Nonlinear. Soft Matter Phys. **66**, 011403-1-4 (2002)
98. Hu, Z.S., Oskam, G., Penn, R.L., Pesika, N., Searson, P.C.: The influence of anion on the coarsening kinetics of ZnO nanoparticles. J. Phys. Chem. B **107**, 3124–3130 (2003)
99. Lifshitz, I.M., Slyozov, V.V.: The kinetics of precipitation from supersaturated solid solutions. J. Phys. Chem. Solids **22**, 35–50 (1961)
100. Wagner, C.: Theorie der alterung von niederschlagen durch umlosen (Ostwald-reifung). Z. Elektrochem. **65**, 581–591 (1961)
101. Huang, F., Zhang, H.Z., Banfield, J.F.: The role of oriented attachment crystal growth in hydrothermal coarsening of nanocrystalline ZnS. J. Phys. Chem. B **107**, 10470–10475 (2003)
102. Huang, F., Zhang, H.Z., Banfield, J.F.: Two-stage crystal-growth kinetics observed during hydrothermal coarsening of nanocrystalline ZnS. Nano Lett. **3**, 373–378 (2003)
103. Peng, X.G., Wickham, J., Alivisatos, A.P.: Kinetics of II-VI and III-V colloidal semiconductor nanocrystal growth: Focusing of size distributions. J. Am. Chem. Soc. **120**, 5343–5344 (1998)
104. Meulenkamp, E.A.: Size dependence of the dissolution of ZnO nanoparticles. J. Phys. Chem. B **102**, 7764–7769 (1998)
105. Greenwood, G.W.: The growth of dispersed precipitates in solutions. Acta Metall. **4**, 243–248 (1956)
106. Kukushkin, S.A., Osipov, A.V.: Crystallization of binary melts and decay of supersaturated solid solutions at the Ostwald ripening stage under non-isothermal conditions. J. Phys. Chem. Solids **56**, 1259–1269 (1995)
107. Liu, Y., Kathan, K., Saad, W., Prudh'omme, R.K.: Ostwald ripening of beta-carotene nanoparticles. Phys. Rev. Lett. **98**, 036102 (2007)
108. Oskam, G., Nellore, A., Penn, R.L., Searson, P.C.: The growth kinetics of TiO$_2$ nanoparticles from titanium(IV) alkoxide at high water/titanium ratio. J. Phys. Chem. B **107**, 1734–1738 (2003)
109. Wong, E.M., Bonevich, J.E., Searson, P.C.: Growth kinetics of nanocrystalline ZnO particles from colloidal suspensions. J. Phys. Chem. B **102**, 7770–7775 (1998)
110. Wong, E.M., Hoertz, P.G., Liang, C.J., Shi, B.M., Meyer, G.J., Searson, P.C.: Influence of organic capping ligands on the growth kinetics of ZnO nanoparticles. Langmuir **17**, 8362–8367 (2001)
111. Hu, Z.S., Oskam, G., Searson, P.C.: Influence of solvent on the growth of ZnO nanoparticles. J. Colloid Interface Sci. **263**, 454–460 (2003)
112. Hu, Z.S., Ramirez, D.J.E., Cervera, B.E.H., Oskam, G., Searson, P.C.: Synthesis of ZnO nanoparticles in 2-propanol by reaction with water. J. Phys. Chem. B **109**, 11209–11214 (2005)
113. Hu, Z.S., Santos, J.F.H., Oskam, G., Searson, P.C.: Influence of the reactant concentrations on the synthesis of ZnO nanoparticles. J. Colloid Interface Sci. **288**, 313–316 (2005)
114. Pesika, N.S., Hu, Z.S., Stebe, K.J., Searson, P.C.: Quenching of growth of ZnO nanoparticles by adsorption of octanethiol. J. Phys. Chem. B **106**, 6985–6990 (2002)

Chapter 4
Oriented Attachment and Mesocrystals

Despite the good applicability of the Ostwald ripening model, recent studies have demonstrated that this mechanism cannot be considered responsible for the growth process in some systems or in nonequilibrium systems [1–9]. The oriented attachment mechanism was proposed as another significant process, which may occur during nanocrystal growth [10–14]. By this mechanism, nanocrystals can grow by the alignment and coalescence of neighboring particles by eliminating a common boundary. The driving force for this mechanism is the decrease in the surface and grain boundaries' free energies. By the localized nature of oriented attachment, the mechanism leads to the formation of nanoparticles with irregular morphologies. Several studies indicate that oriented attachment is very significant, even in the early stages of nanocrystal growth, and may lead to the formation of anisotropic nanostructures in suspensions, such as nanorods, by the consumption of nanoparticles as *building blocks* [15–21]. This mechanism has already been studied theoretically [22–25] and observed experimentally for several years in micrometric metallic systems [26–29].

4.1 A Qualitative Analysis of the Oriented Attachment (OA) Mechanism

A more in-depth analysis of the growth mechanism, such as OA, can shed light on the formation of anisotropic nanostructures as well as mesocrystals. The number of materials obtained by the OA process is growing rapidly [30] and has become an attractive form of processing nanomaterials with anisotropic structures. The OA mechanism originally proposed by Banfield and Penn [3, 5] is a process involving the self-organization of adjacent nanocrystals and coalescence. Basically, we can define two main possible ways to achieve self-organization or mutual orientation of adjacent nanocrystals. One is an effective collision of particles with mutual orientation; the second is coalescence induced by particle rotation. The first situation

E.R. Leite and C. Ribeiro, *Crystallization and Growth of Colloidal Nanocrystals*,
SpringerBriefs in Materials, DOI 10.1007/978-1-4614-1308-0_4,
© Edson Roberto Leite and Caue Ribeiro 2012

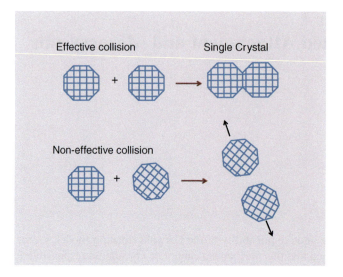

Fig. 4.1 Schematic representation of effective and noneffective collision between particles

must occur in the dispersed colloidal state, in which the number of collisions among particles is high. Particle rotation must be dominant in a weakly flocculated colloidal state, where the interaction among particles is significant, although the nanoparticles still have rotational freedom. Here, one can observe a close relationship between the OA mechanism and the colloidal state [31].

4.1.1   OA in a Dispersed Colloidal State

The dispersed colloidal state is kinetically stable, and the repulsive forces are dominant. In this colloidal state, the OA growth rate is related to the effective collision rate among the nanocrystals in suspension and to the reduction of surface energy aimed at minimizing the area of high-energy facets [18, 32]. Effective collision is when particles produce an irreversible oriented attachment. This occurs only if their orientations at the time of collision achieve a congruent two-dimensional structure at the interface [33–35]. Collision among particles with different crystallographic orientations will not be effective and will not result in irreversible attachment. The schematic in Fig. 4.1 shows the difference between effective and noneffective collision of nanocrystals in a dispersed colloid.

It is well established that oriented attachment can promote the growth of zero-dimensional (0D such as spheres, cubes, or polyhedrons), one-dimensional (1D such as rods and wires), two-dimensional (2D like plates), and three-dimensional self-assembled nanocrystals [36]. In the case of oriented attachment controlled by collision, one will basically observe the growth of 0D (spheres, cubes, or polyhedrons) and 1D structures (rods and wires). In this situation, growth occurs

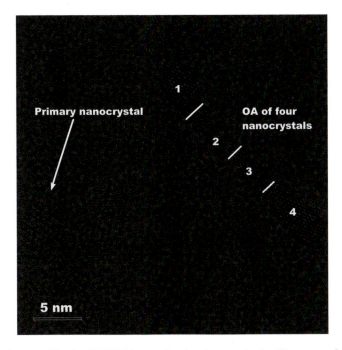

Fig. 4.2 High-magnification HRTEM image showing the growth of a 1D structure from 0D for TiO_2 nanocrystals in the dispersed colloidal state (organic solvent)

through the interaction of 0D with 0D structures, or even through the interaction of 0D with 1D and 1D with 1D structures [37, 38]. The high-resolution transmission electron microscopy (HRTEM) image in Fig. 4.2 illustrates the growth of a 1D structure from a 0D one, forming an anisotropic nanoparticle (a 1D rod formed by two primary particles) of TiO_2 by OA. This process was observed in a low concentration of solids (organic solvent) in a well-dispersed colloidal state.

The OA process that occurs in the dispersed colloidal state can be considered a statistical process controlled by collision frequency. As a consequence of this statistical nature, during the growth process, one can observe isotropic and anisotropic nanocrystals with different growth directions. For instance, in colloidal SnO_2 nanocrystals, the formation of worm-like particles has been observed growing in the [001] and [110] directions, as well as the formation of equiaxed particles [19, 32].

The statistical nature of OA in dispersed colloidal state is highly evident when the nanocrystals are faceted. For instance, the HRTEM image of Fig. 4.3 shows the formation of OA particles with different morphologies, which originate from the sharing of different crystal facets. Today, it can be stated that the OA mechanism controlled by effective collision is related basically to the number and area of the facets in a nanocrystal. In a recent work, Stroppa and coauthors [39] showed a clear correlation between the number and area of facets in faceted nanocrystals of Sb-doped SnO_2 and the OA process controlled by collision in a dispersed system.

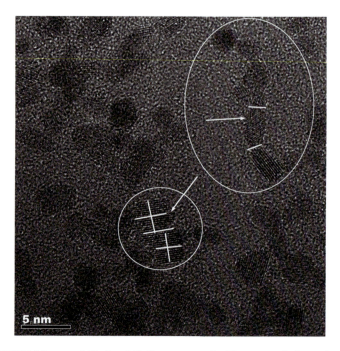

Fig. 4.3 HRTEM image of Sb-doped SnO_2 nanocrystals. The arrows indicate details of the growth process involving different facets

4.1.2   OA in a Weakly Flocculated Colloidal State

In a weakly flocculated colloidal state, the OA process is not controlled by collisions between particles but is dominated by the particle medium- and short-range interaction. In terms of energetic arguments, this is a condition in which the attractive forces between particles, such as van der Waals and structural forces, are higher than repulsive forces (electrostatic and steric, mainly). This energetic condition is normally observed in weakly and strongly flocculated colloidal states. Since rotational freedom between particles is required to achieve crystallographic alignment, the weakly flocculated state is more appropriate.

In this growth condition, OA occurs between primary particles (0D with 0D interaction) resulting in 1D or 2D structures. In the second stage, the 1D and 2D structures can interact, forming 3D structures. In fact, the growth mechanism controlled by OA in this colloidal state follows a hierarchical process and can present mesocrystals as an intermediate state, i.e., a crystal formed by oriented assemblies of nanocrystals [40]. This is a very interesting situation because it enables one to grow mesocrystals with exposed facets not controlled by thermodynamic surface energy, or predicted by the Wulff construction, but controlled kinetically. Consequently, functional materials can be developed with different or even unique chemical, electrical, and magnetic behavior. Figure 4.4 describes a

Fig. 4.4 HRTEM image showing the assembly of TiO$_2$ nanocrystals resulting in the formation of anatase mesocrystals: (**a**) short treatment time in a solvothermal condition, (**b**) long treatment time in the same solvothermal condition. The inset shows the FFT of the region indicated by the *arrow*

typical hierarchical growth process whereby TiO$_2$ anatase nanocrystals assemble in larger agglomerates (Fig. 4.4a) via OA. With increased treatment time, one can observe the formation of well-defined 3D mesocrystals (Fig. 4.4b) with low porosity. The inset of Fig. 4.4b shows the FFT (fast Fourier transform) of the image with reflections typical of a single crystal. It should be noted that this mesocrystal growth process occurs in a weakly flocculated colloidal state, in octyl alcohol.

The statistical model used in a well-dispersed colloidal system cannot explain the mesocrystal growth process depicted in Fig. 4.4. Therefore, a different approach is needed in order to understand this ordered assembly of nanocrystals. By analogy with biological and molecular systems, the 1D, 2D, and 3D morphologies attained by OA can be described by noncovalent interactions, i.e., by interactions with much weaker energy than that of covalent bonds (~100–150 k_BT), in the range of 0.1–10 k_bT. In the weakly flocculated state, the repulsive forces may be negligible; hence, the van der Waals forces (vdW) will be dominant.

A direct correlation between dipole interactions and nanocrystal and mesocrystal growth has been demonstrated in the literature [15, 41]. In a classical work [15], Giersig and coauthors described the spontaneous organization of single CdTe nanocrystals into 1D nanostructures (nanowires), attributing this spontaneous interaction to dipole–dipole attraction. In their work, they estimated the potential attractive energy to be ~10 kJ/mol. One of the key steps in the preparation of the 1D nanostructure was the removal of excessive stabilization through the precipitation of nanocrystals by the addition of methanol. This experiment was performed at room temperature. In a recent study, Kotov and coauthors [41] showed that the dipole moment, a small positive charge, and hydrophobic attraction are the driving forces for the self-organization of CdTe into 2D structures resembling the assembly of surface layer (S-layer) proteins [41].

Leite and coworkers [42] demonstrated that the use of microwave (MW) heating during the growth process reduces the treatment time required to obtain anisotropic nanostructures. They reported that OA is the dominant mechanism responsible for the growth process, implying that it could be possible to introduce MW irradiation as a parameter to control the anisotropic growth of crystals [42]. The authors attributed the shorter treatment time to the increase of the effective collision rate, i.e., the presence of attractive forces. A plausible explanation for the increase of the effective collision rate under microwave irradiation is the increase in the collision cross section (σ) of the particle. In a previous article [34], Leite's group made a detailed kinetic analysis of the OA mechanism, in which it was assumed that a noninteractive collision occurred with a collision cross section (σ), $\sigma = 3r$, where r is the particle's radius. Under microwave irradiation, it was assumed that an interactive collision occurred which introduced a steric factor (P), expressing the interactive cross section $(\sigma*)$ as a multiple of σ:

$$\sigma* = P\sigma, \tag{4.1}$$

where $P > 1$. The origin of the interactive collision, and hence, of the steric factor, is assumed to be the instantaneous dipole moment generated in the nanocrystal by the electromagnetic field induced by microwave radiation, i.e., induced dipole-induced dipole interaction, or LD forces.

From the above discussion, it is clear that the colloidal state strongly influences the OA growth process. For a better description of this dependence, one can use an energetic argument to define when OA will be controlled by the collision rate or by interacting nanoparticles. For a simplified description of the interparticle interaction and thermal fluctuation, we will consider only the kinetic energy (KE) of an ideal gas and the attractive potential. At first glance, these considerations may be too drastic for a real system; however, our goal is to come up with a description that allows for a qualitative and phenomenological analysis of the OA process.

We can estimate the KE of one particle, using the following equation:

$$KE = 1.5k_B T, \tag{4.2}$$

where k_B is the Boltzmann constant.

As can be seen, the KE is linearly dependent on the temperature (T). Now, analyzing the attractive potential and considering the thermal fluctuation, one can use Eq. 2.20 described in the Chap. 2, for instance, which describes an attractive dipole–charge interaction. As we can see, $\langle V \rangle_{d-c}$ is inversely proportional to temperature. Didactically, the KE and the modulus of $\langle V \rangle_{d-c}$ can be plotted as a function of T (see Fig. 4.5), where one observes a crossover point. At this point, a critical temperature, T_c, can be defined. This temperature defines when the thermal fluctuation exceeds the attractive potential energy. For a condition in which $T > T_c$, the KE will predominate (the colloidal dispersion will be in a kinetically stable condition), and one can postulate that the statistical and collision rate approach will control the OA growth process. Now, considering $T < T_c$, the attractive potential energy will be dominant, and the OA process will be controlled by the particle medium and short range interactions. This energetic condition generally leads to a weakly flocculated colloidal state.

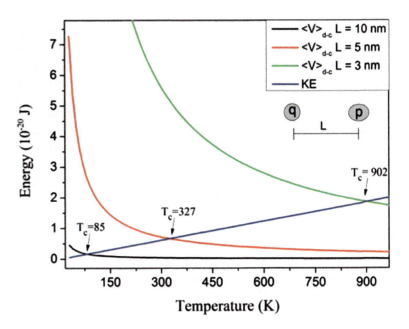

Fig. 4.5 KE and the modulus of $\langle V \rangle_{d-c}$ as function of temperature, considering different L values (the distance between molecules or nanocrystals)

Analyzing Fig. 4.5, one can also see that different interparticle distances (L) result in different T_c. In fact, T_c is a term that must be dependent on other parameters besides L. T_c can be defined as:

$$T_c = (cpq)/\sqrt{1.5k_BL^2}. \tag{4.3}$$

This is an interesting equation. The terms p and q are parameters related to the interactions and can be related easily to the steric factor P described in Eq. 4.1. On the other hand, the steric factor is related to the interactive cross section (σ^*) and to the collision cross section (σ). In a first approximation, the parameter L, or in a system with N particles, the mean interparticle distance $\langle L \rangle$, should be inversely proportional to the specimen concentration ($[A]$). Based on this consideration, Eq. 4.3 can be rewritten to obtain a more general equation:

$$T_c = C''\sigma^*[A]^{n'}/r, \tag{4.4}$$

where C'' is an empirical constant and n' is a power-law exponent. This relation can help us determine the parameters that are important in controlling the nanocrystal growth process or even in controlling the formation of mesocrystals (a mesoscopically structured crystal). In addition to the concentration, the solution chemistry is an important parameter and is represented by σ^*. A dependence on nanocrystal size (r) is also observed.

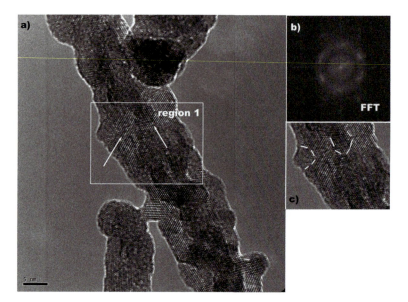

Fig. 4.6 HRTEM image of the GCO growth at 40°C: (**a**) High-magnification HRTEM image, (**b**) FFT of region 1 indicated in Fig. 4.11a, (**c**) reconstructed lattice image of region 1

This qualitative analysis is consistent with experimental results reported recently by Leite and coauthors [31] for the synthesis of gadolinium-doped cerium oxide (GCO) nanorods. As shown by scanning electron microscopy (SEM) characterization, decreasing the growth temperature increases the yield of nanorods. At 40°C, a large number of nanorods were observed. Upon changing the temperature, differences in rod size and concentration were reported. At 200°C, one can no longer see the formation of rods but only equiaxed nanoparticles. A HRTEM characterization of GCO nanorod, grown at 40°C, shows features typical of the OA process (Fig. 4.6a, b). This figure shows the presence of a mesocrystal-like structure, with the clear presence of a boundary between oriented primary particles, as indicated in the reconstructed lattice image of Fig. 4.6c. One can also see an irregular rod surface. Figure 4.7 shows HRTEM images of materials treated at a higher temperature. Upon increasing the growth temperature to 130°C, the presence of a grain boundary is no longer evident, indicating the formation of single crystalline nanorods with a smooth surface (self-recrystallization process) (Fig. 4.7a). In fact, this observation indicates that the temperature is an important parameter to promote grain boundary elimination and surface reconstruction, suggesting that a thermally activated process, such as diffusion, controls both phenomena. High-magnification HRTEM images of the GCO growth at 200°C (Fig. 4.7b) shows features typical of the OA process induced by effective collision between nanocrystals.

Based on these experimental results, we can postulate that increasing the temperature causes the thermal fluctuation to increase (mainly the KE), resulting in a statistical growth process controlled by the collision rate. Low temperatures favor the predominance of attractive interactions, leading to the formation of nanorods

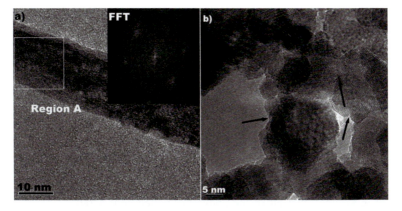

Fig. 4.7 High-magnification HRTEM image of the GCO: (**a**) growth at 130°C. The inset shows a FFT of region a, (**b**) growth at 200°C. The *arrows* indicate typical features of the OA process

(mesocrystals). This analysis is entirely consistent with the idea of a transition temperature between a growth regime controlled by interaction and a regime controlled by statistical collision.

The next challenge in the study of OA is the proposal of a quantitative kinetic model to describe this growth mechanism in a weakly flocculated state. Of course, this is not a simple endeavor since such a model must include the attractive potential between nanocrystals. In fact, the first step in this direction has already been taken. In his paper about a kinetic model, Lee Penn [35] proposed to correlate the rate constant with an attractive potential between nanoparticles. This proposal may be a good starting point, but the model requires further improvement in order to obtain a representative kinetic equation.

4.2 Quantitative Description of the OA Mechanism

In recent years, several models have been proposed to explain or describe the OA growth mechanism [34, 35], and most of them are applied in a condition of well-dispersed colloidal state. Below is a detailed description of a model in which the nanocrystals or nanoparticles are considered as molecules, and the reaction is controlled by collision between nanoparticles.

4.2.1 Oriented Attachment in the Dispersed Colloidal State: A Quantitative Description

To explain the growth behavior in SnO_2 colloidal suspensions, Ribeiro et al. [43] proposed that coalescence in nonagglomerated suspensions may also occur when particles with similar crystallographic orientations (or with slight differences) collide. This mechanism is based on the assumption that nanoparticles dispersed in a liquid medium present a very high degree of freedom for rotation and translation motions. Hence, in suspensions where agglomeration does not take place, growth by means of oriented collisions should be more effective than by surface mechanisms (i.e., coalescence induced by relative rotations between particles in contact). Dispersed nanoparticles should present a high velocity due to the Brownian motion. Hence, it is expected that nanoparticles in suspension present a high frequency of collisions. Therefore, growth by coalescence may be interpreted statistically since collisions may be considered effective (i.e., leading to coalescence) or ineffective (i.e., *elastic* event). This mechanism is similar to the Smoluchowsky coagulation model [44–46] used extensively to explain polycrystalline colloidal growth and aggregation mechanisms in suspension.

If it is assumed that all of the above-mentioned considerations are valid, the coalescence of two particles in suspension may be interpreted through the following chemical Eq. 4.5:

$$A + A \rightarrow B, \tag{4.5}$$

where A is a primary nanoparticle and B is the product of coalescence of two nanoparticles. The idea of a chemical reaction where the nanoparticles act as molecules was also proposed by Huang et al. [47–49], Penn [35], and Drews et al. [50, 51] using other approximations. Also, some researchers used this interpretation to gain a more thorough understanding of the experimental growth behavior in colloidal suspensions of several systems, such as ZnS [47, 52], SnO_2 [19, 53], TiO_2 [54], Nb_2O_5 [55], and others.

The first step is to define if the collision frequency of dispersed nanoparticles may be significant to the process. This can be evaluated by assuming that Brownian motion in dilute suspensions may be described by Maxwell–Boltzmann statistics. In this model, the frequency evaluation is done in analogy to the kinetics of gas molecules. The collision frequency for a single particle is given as a function of the mean velocity, ϖ by Eq. 4.6:

$$z = (\sqrt{2}\pi D^2 \varpi N)/V, \tag{4.6}$$

where D is the particle diameter, N is the total number of particles, and V is the total volume occupied by the system. The viscous force is given by $\mu \pi^2 \varpi D^2$, where μ is the viscosity, and it is negligible for systems composed of low-viscosity fluids, as in the case of nanoparticles dispersed in water. Hence, the mean velocity of the dispersed particles may be estimated by the equipartition theorem (4.7):

$$\varpi = \sqrt{(3k_B T/m)}. \tag{4.7}$$

The mass of a spherical nanoparticle with a radius of 2–5 nm and a density of 3–10 g/cm^3 is on the order of 1×10^{-8} g. At room temperature, this nanoparticle presents a mean velocity of ~1.1 m/s, which is a very high value if particle size is considered. In the study reported by Ribeiro et al. [34], the SnO_2 nanoparticle concentration (for dilute suspensions) was estimated in the order of 1×10^{18} particles per liter.

If these parameters are inserted in Eq. 4.6, the collision frequency is estimated to be ~240 collisions/s for each particle. Although this value may not be considered precise, it still indicates that the number of (total) collisions may be indeed significant. It can be observed that the nanoparticle collision frequency is much lower than the value expected for gas molecules, which is obviously due to mass effects. Moreover, it is also important to note that collisions are only effective if particles with the same crystallographic orientation collide.

Returning to Eq. 4.5, the *rate of reaction* (v) of the oriented attachment mechanism may be given by (4.8):

$$v = -(1/2)d[A]/dt = k[A]^2, \qquad (4.8)$$

where $[A]$ is the concentration of primary (i.e., uncoalesced) particles (particles/volume), t the time, and k is the reaction constant. Assuming that the reaction occurs in a single step, $[A]$ is defined in terms of the initial concentration $[A]_0$ by Eq. 4.9:

$$[A] = [A]_0/(1 + 2k[A]_0 t). \qquad (4.9)$$

However, this result is applicable only if particles are in direct contact. In suspensions, particles need to achieve an equilibrium condition, which is provided by collisions, in order to form *complexes* – i.e., two particles in contact – that may coalesce. This process may be described as a two-step reaction, as shown in Eqs. 4.10 and 4.11:

$$A + A \Leftrightarrow AA \quad \text{(rate constants } k_1, k_1'), \qquad (4.10)$$

$$AA \rightarrow B \quad \text{(rate constant } k_2) \qquad (4.11)$$

Therefore, the kinetics of this process is described by three reaction rates, $v_{1\text{forward}}$ and $v_{1\text{reverse}}$ (Eqs. 4.12 and 4.13), which corresponds to the *complex* formation equilibrium reaction, and v_2, which is related to the coalescence event (4.14):

$$v_{1\text{ forward}} = k_1[A]^2, \qquad (4.12)$$

$$v_{1\text{ reverse}} = k_1'[AA], \qquad (4.13)$$

$$v_2 = k_2[AA]. \qquad (4.14)$$

Assuming that the *complex* concentration is in the steady state, i.e., $d[AA]/dt = 0$ (4.15), it can be written as (4.16):

$$d[AA]/dt = k_1[A]^2 - k_1'[AA] - k_2[AA] = 0, \qquad (4.15)$$

$$[AA] = k_1[A]^2/(k_1' + k_2). \qquad (4.16)$$

The rate of formation of coalesced particles B (Eq. 4.17) is obtained by Eqs. 4.14 and 4.16:

$$d[B]/dt = -(1/2)(d[A]/dt)(k_2 \cdot k_1/k_1' + k_2)[A]^2. \qquad (4.17)$$

This equation can be solved similarly to the solution of Eq. 4.9, where $(k_2 k_1/k_1' + k_2)$ can be interpreted as the rate constant k_T of the total reaction. According to the initial proposition, oriented collision-induced coalescence should be a very fast process when compared to mechanisms of coalescence induced by particle rotation. Therefore, it is clear that the attachment will be dominated by the first step (i.e., collision process), and it can be assumed that $k_T \approx k_1$.

The total particle flux J_T through a stationary spherical particle is given by Eq. 4.18:

$$J_T = 4\pi r^2 \cdot J, \qquad (4.18)$$

where J is the flux around a particle (A) with surface area $4\pi r^2$. As defined by Fick's first law (4.19):

$$J = D_A \cdot d[A]/dx, \qquad (4.19)$$

where D_A is the diffusion coefficient of A and x is the distance. This term is obtained in analogy with atomic diffusion, and it actually describes the diffusion process of nanoparticles. The concentration of primary particles $[A]$ may be defined in terms of the total particle flux (4.20) by integrating Eq. 4.19:

$$[A]_x = [A] - (J_T/4\pi D_A x). \qquad (4.20)$$

During collisions, whenever a particle is within a distance of $2r$ from the surface of another particle A, the formation of the *complex AA* occurs, as shown in Fig. 4.8. Hence, in this situation, the concentration of primary particles around the collision site can be considered equal to zero (i.e., $[A]_x = 0$), and J_T can be given as a function of $[A]$ and $R = 3r$ by (4.21):

$$J_T = 12\pi D_A r[A]. \qquad (4.21)$$

Fig. 4.8 Model of the contact of two particles *A* forming a *complex AA*. The *dashed line* corresponds to the collision cross section

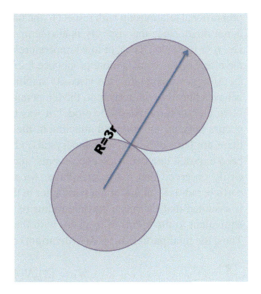

The number of particles in suspension is defined as $[A]NV$, where N is Avogadro's number and V is the total volume of the suspension. Therefore, the overall flux of particles can be defined as $J_T[A]NV$ since particles are not really stationary. Since the overall flux is time-dependent, the following approximation (Eq. 4.22) can be used:

$$d[AA]/dt = (6\pi r D_A N)[A]^2. \tag{4.22}$$

By comparison with Eq. 4.12, the term in parenthesis can be interpreted as the reaction rate constant k_1, as follows (Eq. 4.23):

$$k_1 = 6\pi r D_A N. \tag{4.23}$$

This result indicates that the particle radius affects the reaction rate. However, this dependence can be reevaluated if the diffusion constant is assumed to be equivalent to the definition given by the Stokes–Einstein Eq. 4.24:

$$D = k_B T / (6\pi \eta r), \tag{4.24}$$

where η is the viscosity of the liquid medium. Hence, the rate constant k_1 may be defined as (4.25):

$$k_1 = N k_B T / \eta. \tag{4.25}$$

This very simple result shows that the viscosity of the liquid medium plays an important role in the growth by particle coalescence and is governed by an inverse proportional relationship with respect to the rate constant. The temperature dependence is not direct since the viscosity is also temperature-dependent in an Arrhenius-form, i.e., $\eta = \eta_o e^{Ea/Nk_BT}$. Thus, at low temperatures (near the freezing point of the solvent), the effect of viscosity is more pronounced, inhibiting the increase of the rate constant k_2. At high temperatures, the viscosity stabilizes to a nearly constant value; hence, k_2 behaves linearly with respect to the temperature.

In the development of the model, it was initially considered that particle growth occurs only by the oriented attachment mechanism. Therefore, it is assumed that there are only two types of nanocrystals: (1) primary particles, A, which have not been exposed to any coalescence events, and (2) coalesced particles, B. Hence, it is easily observed that B particles have twice the volume of A particles. The mean particle radius is an important parameter in the evaluation of nanoparticle growth. It is assumed that the mean particle radius of a coalesced particle can be considered equivalent to the radius of a sphere with the same volume (i.e., equivalent radius). Thus, the total particle mass M_T (invariant) can be described by Eqs. 4.26 and 4.27:

$$M_T = [A]_0 N (4/3) \pi r_i^3, \tag{4.26}$$

$$M_T = ([A] + [B]) N (4/3) \pi r_{eq}^3, \tag{4.27}$$

where r_i is the initial mean particle radius and r_{eq} is the equivalent radius at a time t. From Eq. 4.5, the total number of particles can be expressed as $[A]_0 = [A] + 2[B]$. By comparing this relationship with the expressions above, one can see that the equivalent diameter depends on $[A]$ as follows:

$$[A]_0 r_i^3 = ([A] + [A]_0/2) r_{eq}^3. \tag{4.28}$$

By inserting Eqs. 4.9 and 4.25 into Eq. 4.28, it is possible to write the following relation:

$$r_{eq}^3 - r_i^3 = \{[(Nk_BT/\eta)[A]_0 t]/[1 + (Nk_BT/\eta)[A]_0 t]\} r_i^3. \tag{4.29}$$

Since all the terms except t are constant, one can see that this equation behaves as a function of the type $y = x/(1 + x)$. This behavior is slightly different from the one expected for the Ostwald ripening mechanism. For very long periods of time, Eq. 4.29 stabilizes at a constant value that corresponds to the moment when all the primary particles have undergone coalescence. This interpretation considers only the first stage of coalescence, in which a single coalescence event occurs for each particle. However, for long periods of time, events such as $A + B \rightarrow C$ (attachment of previously coalesced particles) may also occur. As a matter of fact, it is highly improbable that each particle coalesces only once.

The model showed good applicability for SnO_2 nanoparticles in hydrothermal conditions [6, 34], since this system clearly grows only by this mechanism. However, in other systems such as CdSe or InAs [34], the equation was not well fitted

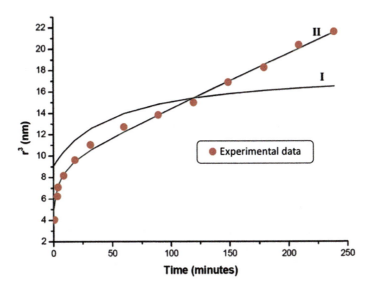

Fig. 4.9 Comparison of Eqs. 4.29 (I) and 4.30 (II) using Pesika's ZnO data [57]

to the experimental data. The deviations in the data may be explained, in part, by the presence of the Ostwald ripening mechanism associated to the OA mechanism in these nanomaterials. Note that it is interesting to analyze the results of particle growth as a function of r^3, since this enables an evaluation of the growth processes in terms of mean particle volumes. Consequently, it is reasonable to assume that the contributions of Ostwald ripening and oriented attachment mechanisms to particle growth are correlated. Considering that coalescence predominates in the initial stages and that Ostwald ripening may take place later, one can assume that the initial radius of coarsening can be given by Eq. 4.29, i.e., $r_{eq} = r_{initial}$. Thus, particle growth may be described by:

$$r^3 - r_{eq}^3 = [(8\gamma V_m^2 c_\infty)/(54a\pi\eta N)] \cdot t, \qquad (4.30)$$

where γ is the surface energy, c_∞ is the bulk solubility, V_m is the molar volume, and a is the radius of the solvated ion. The assumptions used to obtain the coefficient $n = 3$ in Ostwald ripening are present in this equation, indicating a time-activated process [56]. Figure 4.9 compares Eqs. 4.29 and 4.30 applied to the experimental data reported by Pesika et al. [57], where the deviations over large time intervals are clearly visible. Similar profiles were observed in nanocrystalline ZnO and TiO_2 growth [58–63].

Ribeiro and coauthors [64] introduced a slight modification into the previously proposed Oriented Attachment model, based on the classical model for stepwise polymerization [65] described by Flory [66, 67]. In this case, the *reaction* may be interpreted as the junction of two active surfaces. For the sake of simplicity, it is considered that a primary particle in suspension (identified as A in Eq. 4.5) behaves

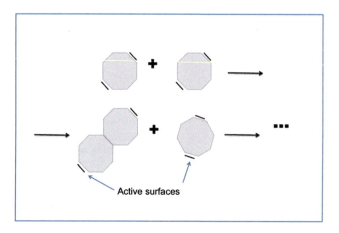

Fig. 4.10 Scheme of the oriented attachment model represented by Eq. 4.31

like an *S-A-S* molecule, where *S* represents an active surface and *A* corresponds to the *body* of the particle – as shown schematically in Fig. 4.10. It is assumed that the coalescence events will follow the chemical equation below:

$$S - A - S + S - A - S \rightarrow S - AA - S \rightarrow S - (A)_x - S. \tag{4.31}$$

The $S - (A)_x - S$ particle is defined as a single particle that has undergone a coalescence sequence of *x* primary particles. Since the active center in all the chemical equations is *S*, and assuming *k* is constant throughout the process, the consumption rate of active surfaces is then expressed by the following equation:

$$v = dS/dt = -k[S]^2. \tag{4.32}$$

The second-order rate law can be solved easily by integrating the Eq. 4.32:

$$[S] = [S]_0/(1 + k[S]_0 t), \tag{4.33}$$

where S_0 is the initial concentration of active surfaces. According to the initial assumptions, $S_0 = 2A_0$, where A_0 is the initial concentration of particles.

The probability *P* of a particular link occurring between two particles is defined in terms of the total amount of active surfaces reacted at time *t*:

$$P = ([S]_0 - [S])/[S]_0. \tag{4.34}$$

According to this definition, the probability of the inexistence of links between two particles is $1 - P$. In a system where *x* configurations are possible, the number of links in a single structure is $x - 1$, and each structure will be left with two potential links (i.e., two active surfaces) (see Fig. 4.10). Hence, the probability of

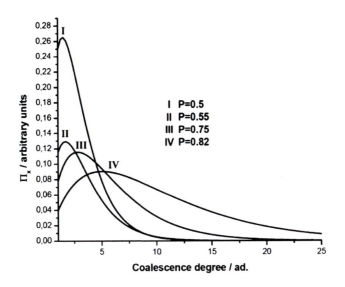

Fig. 4.11 Examples of distribution of configurations or coalescence degrees for different values of P

the existence of a particular configuration is $P^{x-1} \cdot (1 - P)^2$. Therefore, the probability of the existence of any configuration x is given by:

$$\prod_x = x \cdot P^{x-1} \cdot (1 - P)^2. \tag{4.35}$$

Equation 4.35 can be interpreted as the probability distribution for particle configurations. Figure 4.11 shows examples of such distributions of configurations, obtained by assuming consumptions of 50%, 55%, 75%, and 82% of the total amount of reactants. The configurations can be represented as a variable, the *coalescence degree*, i.e., the number of particles required to form a single coalesced particle. The plotted curves depict an important feature: The average size is not an adequate parameter to represent linear growth, especially in the initial stages, since the curves are not symmetrical. In this sense, the most probable configuration (i.e., the most probable degree of coalescence) should represent the growth more adequately. Since the most probable configuration is given by $d\Pi_x/dx = 0$, one has:

$$d\Pi_x/dx = (1 - P)^2(P^{x-1} + xP^{x-1}\ln P). \tag{4.36}$$

The solutions are $x_{min} = \infty$ and $x_{max} = -1/\ln P$, where $0 < P < 1$. Substituting P (Eq. 4.34) in the second solution, one has:

$$x_{max} = -1/\ln\{([S]_0 - [S])/[S]_0\}. \tag{4.37}$$

Replacing the value of $[S]$ as defined in Eq. 4.33, one has:

$$x_{max} = -1/\ln(k[S]_0 t/1 + k[S]_0 t). \qquad (4.38)$$

The solution x_{max} (Eq. 4.38) gives the most probable configuration for a given t and represents a more consistent interpretation of the growth behavior of aniso-tropic nanocrystals: The initial stages are determined by a strong logarithmic behavior followed by a linear-type behavior for high consumption of reactants, since $-1/\ln P \approx 1/(1-P)$ for P near 1, i.e., $x_{max} \approx 1 + k[S]_0 t$ in this condition. At this limit, the most probable configuration is equal to the average configuration.

The existence probability, as described by Eq. 4.35, can be interpreted as the distribution of configurations or as distributions of particle sizes or weights. Ribeiro and coauthors [64] applied Eq. 4.35 to the distribution data of a SnO_2 colloidal sample subjected to hydrothermal processing for 24 h at 200°C. The result showed a significant congruence between the experimental data and the fitted data, where the value of P obtained (0.76) is highly representative, since this value, when applied to Eq. 4.34, results in an average degree of coalescence of 4.2 (i.e., on average, a particle in the system is composed of 4.2 primary particles).

Although an approximate calculation indicated that k assumes a constant value during the overall process (see Eqs. 4.32 and 4.33), this can be a reasonable assumption for extended reaction times. On the other hand, with shorter times, the mobility of coalesced particles is strongly affected in relation to uncoalesced particles. The dependence of k is estimated based on the size of the particles involved in the reaction, using the kinetic theory of gases. According to this approach, the collision density of pairs in random systems is given by the collision cross section of the two particles involved (defined as $\pi(R_1 + R_2)^2$, where R_1 and R_2 are the radii of the two particles), and by the reduced mass of the two particles $(1/m_1 + 1/m_2)$, as follows:

$$k = P_{steric} \pi (R_1 + R_2)^2 (1/m_1 + 1/m_2)^{1/2} (8k_B T N_A^2)^{1/2} \cdot e^{-E_a/RT}, \qquad (4.39)$$

where P_{steric} is the steric factor (the probability of a successful collision), k_B is Boltzmann's constant, T is the temperature, N_A is Avogadro's constant, E_a is the activation energy, and R is the universal gas constant. Since it was also assumed that the particles formed are linear, the probability p of a collision at the extremities is determined by the gyration ratio of the anisotropic particle, i.e.:

$$p \approx 2.2R/(\pi n R) = 4/(\pi n), \qquad (4.40)$$

where n is the number of primary particles in the resulting particle. Now, considering two particles composed of n and m primary particles, as illustrated in Fig. 4.12, the probability of a successful collision between two particles (P_{steric}) is given by:

$$P_{steric} = (4^2/\pi^2)(1/(n \cdot m)). \qquad (4.41)$$

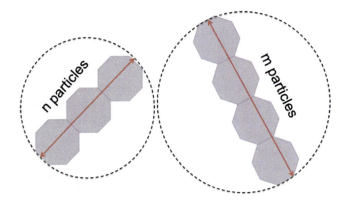

Fig. 4.12 Scheme of the collision between two particles formed by n and m primary particles

Similarly, the collision cross section will be determined by the gyration ratio of the anisotropic particles. Finally, the masses of the particles are equal to $n \cdot M_{initial}$ and $m \cdot M_{initial}$, where $M_{initial}$ is the mass of the primary particle. Substituting this whole expression in Eq. 4.39, and regrouping all the constant terms into a single constant, k_0, one has:

$$k = k_0[1/(n \cdot m)](n + m)^2(1/n + 1/m)^{1/2}. \tag{4.42}$$

Considering the specific case of $n = m$, one has:

$$k = 4\sqrt{2}k_0(1/n). \tag{4.43}$$

This result (Eq. 4.43) shows that the value k cannot be stated as being constant in the entire range of the reaction. Because k behaves as $y = 1/x$, this value should be considered constant only during extended reaction times. Considering that n is equal to $[S]_0/[S]$ and replacing this value in Eqs. 4.43 and 4.32, one obtains a rate law expressed as:

$$d[S]/dt = k_0'[S]^3, \tag{4.44}$$

where $k_0' = (4\sqrt{2}k_0)/[S]_0$. The integration of the above expression results in:

$$[S] = [S]_0/\{(1 + 8\sqrt{2}[S]_0k_0t)^{1/2}\}. \tag{4.45}$$

Applying this result to the definition of P and x_{max} (Eqs. 4.34 and 4.37), one obtains a general form of the time dependence of x_{max}. However, the exponent 1/2 in Eq. 4.45 is obtained through the approximation $n = m$, i.e., the collision of two particles of equal coalescence number. Based on the above, it is suggested that a more coherent approach is to assume the existence of an exponent α, where $0 < \alpha < 1$, in which case the general expression takes on the following form:

$$x_{max} = -1/\{\alpha \ln(8\sqrt{2}k_0[S]_0 t/(1 + 8\sqrt{2}k_0[S]_0 t))\}. \tag{4.46}$$

This final expression was fitted with good agreement to experiments performed by processing SnO_2 hydrothermal colloids reported by Lee [53, 64] and using the experimental data of Lee Penn and Banfield [5] for the case of anatase growth in hydrothermal experiments.

One condition assumed in the model (the existence of only two active surfaces) cannot be guaranteed since the growth process is determined statistically. In this context, some of the systems reported in the literature, in which growth occurs in the presence of surfactants adhering to specific surfaces of the primary particle, are better suited to the model, for example, the growth of TiO_2-anatase nanorods reported by Jun and coworkers [68] or the similar experiment performed by Cozzoli et al. [69] and Polleux et al. [17, 18]. Some authors have reported that the formation of highly anisotropic structures (e.g., nanowires) occurs by oriented attachment and depends on the existence of a mechanism for organized agglomeration [6, 8, 9, 19]. Tang and coworkers [15] discussed the formation of CdTe nanowires from *building-block* nanoparticles, whereby the shapes achieved were brought about by a spontaneous orientation between the nanoparticles in response to dipole interactions in the liquid medium. In this route, the nanoparticles are covered with a surfactant, which is gradually eliminated from the surfaces during consecutive coalescence events, finally resulting in a nanowire. Under such conditions, the active surfaces responsible for the oriented attachment are not covered by surfactants. Cho and coworkers [70] used the same argument to explain the formation of PbSe nanowires and nanorings in solution. In this work, the authors attribute the alignment explicitly to dipole interactions as a first step in the process, with oriented attachment as the subsequent step. These types of phenomena can be approximated easily to the idea of two active surfaces, as proposed. In a recent work, Kumacheva and coauthors [71] proposed a similar kinetic model to describe the self-organization of nanoparticles. In this work, they reported on the marked similarity between the self-assembly of metal nano-particles and reaction-controlled step-growth polymerization.

As mentioned earlier, experimental results for nanocrystal synthesis or growth reflect, to a certain extent, a model or an intermediate situation, as in the growth of agglomerated nanocrystals. In SnO_2 nanocrystals, Lee and coauthors [19] showed that agglomeration can play an important role in the growth of SnO_2 nanoribbons. Under dispersed conditions, anisotropic particles are formed by successive collisions without particle rotation. Here, anisotropic growth is statistical and can lead to various particle shapes. Under agglomerated conditions, alignment by grain rotation may occur if the particles are in contact. However, since the crystallo-graphic alignment can be satisfied at any point of the particle's surface, the final particle can result from several attachment events along the same surface, resulting in an uncontrollable shape. If the agglomerating agent can cause a significant steric effect, as in the case of organic acids, the growth is hindered by the reduction of the

Fig. 4.13 Low-magnification image of ZrO_2 nanocrystals, showing a quasi-monodisperse distribution, synthesized according the procedure of Garnweitner et al. [73]. The inset shows a high-magnification HRTEM image of a well-crystallized ZrO_2 nanocrystal

Brownian motion, affecting the effective collision mechanism, and by the suppression of particle-particle contacts affecting the driving force for rotation, i.e., the net torque resulting from the interfacial energies. As the above-mentioned authors observed, growth in agglomerated conditions is much faster than in dispersed conditions, however entirely uncontrollable. These hypotheses lead to the use of selective surfactants, i.e., compounds having a high affinity with some crystallographic planes of the nanoparticle. In general, this hypothesis has been proposed in several studies on the synthesis of nanocrystals in organic media [72, 73], with the authors postulating that the medium acts as a surfactant of the as-formed nanoparticles, hindering growth. In fact, in all the cited studies, the authors obtained almost monodispersed nanoparticles, suggesting that size was controlled mainly by nucleation and reaction growth events. This is illustrated in the low-magnification HRTEM image of ZrO_2 nanocrystals synthesized by a nonhydrolytic route, in Fig. 4.13.

References

1. Pacholski, C., Kornowski, A., Weller, H.: Self-assembly of ZnO: From nanodots, to nanorods. Angew. Chem. Int. Ed. **41**, 1188 (2002)
2. Penn, R.L., Oskam, G., Strathmann, T.J., Searson, P.C., Stone, A.T., Veblen, D.R.: Epitaxial assembly in aged colloids. J. Phys. Chem. B **105**, 2177–2182 (2001)
3. Banfield, J.F., Welch, S.A., Zhang, H.Z., Ebert, T.T., Penn, R.L.: Aggregation-based crystal growth and microstructure development in natural iron oxyhydroxide biomineralization products. Science **289**, 751–754 (2000)

4. Peng, X.G., Manna, L., Yang, W.D., Wickham, J., Scher, E., Kadavanich, A., Alivisatos, A.P.: Shape control of CdSe nanocrystals. Nature **404**, 59–61 (2000)
5. Penn, R.L., Banfield, J.F.: Morphology development and crystal growth in nanocrystalline aggregates under hydrothermal conditions: Insights from titania. Geochim. Cosmochim. Acta **63**, 1549–1557 (1999)
6. Lee, E.J.H., Ribeiro, C., Longo, E., Leite, E.R.: Growth kinetics of tin oxide nanocrystal in colloidal suspensions under hydrothermal conditions. Chem. Phys. **328**, 229–235 (2006)
7. Colfen, H., Mann, S.: Higher-order organization by mesoscale self-assembly and transformation of hybrid nanostructures. Angew. Chem. Int. Ed. **42**, 2350–2365 (2003)
8. Colfen, H., Antonietti, M.: Mesocrystals: Inorganic superstructures made by highly parallel crystallization and controlled alignment. Angew. Chem. Int. Ed. **44**, 5576–5591 (2005)
9. Niederberger, M., Colfen, H.: Oriented attachment and mesocrystals: Non-classical crystallization mechanisms based on nanoparticle assembly. Phys. Chem. Chem. Phys. **8**, 3271–3287 (2006)
10. Nespolo, M., Ferraris, G.: Applied geminography – symmetry analysis of twinned crystals and definition of twinning by reticular polyholohedry. Acta Crystallogr. Sect. A: Found. Crystallogr. **60**, 89–95 (2004)
11. Nespolo, M., Ferraris, G., Durovic, S., Takeuchi, Y.: Twins vs. modular crystal structures. Z. Kristallogr. **219**, 773–778 (2004)
12. Nespolo, M., Ferraris, G.: The oriented attachment mechanism in the formation of twins - a survey. Eur. J. Mineral. **16**, 401–406 (2004)
13. Nespolo, M., Ferraris, G.: Hybrid twinning - a cooperative type of oriented crystal association. Z. Kristallogr. **220**, 317–323 (2005)
14. Penn, R.L., Banfield, J.F.: Oriented attachment and growth, twinning, polytypism, and formation of metastable phases: insights from nanocrystalline TiO_2. Am. Mineral. **83**, 1077–1082 (1998)
15. Tang, Z.Y., Kotov, N.A., Giersig, M.: Spontaneous organization of single CdTe nanoparticles into luminescent nanowires. Science **297**, 237–240 (2002)
16. Chushkin, Y., Ulmeanu, M., Luby, S., Majkova, E., Kostic, I., Klang, P., Holy, V., Bochnicek, Z., Giersig, M., Hilgendorff, M., Metzger, T.H.: Structural study of self-assembled Co nanoparticles. J. Appl. Phys. **94**, 7743–7748 (2003)
17. Polleux, J., Pinna, N., Antonietti, M., Niederberger, M.: Ligand-directed assembly of preformed titania nanocrystals into highly anisotropic nanostructures. Adv. Mater. **16**, 436 (2004)
18. Polleux, J., Pinna, N., Antonietti, M., Hess, C., Wild, U., Schlogl, R., Niederberger, M.: Ligand functionality as a versatile tool to control the assembly behavior of preformed titania nanocrystals. Chem. Eur. J. **11**, 3541–3551 (2005)
19. Lee, E.J.H., Ribeiro, C., Longo, E., Leite, E.R.: Oriented attachment: An effective mechanism in the formation of anisotropic nanocrystals. J. Phys. Chem. B **109**, 20842–20846 (2005)
20. Ribeiro, C., Vila, C., Stroppa, D.B., Bettini, J., Mastelaro, V.R., Longo, E., Leite, E.R.: Anisotropic growth of oxide nanocrystals: insights into the rutile TiO_2 phase. J. Phys. Chem. C **111**, 5871–5875 (2007)
21. Ribeiro, C., Vila, C., Matos, J.M.E., Bettini, J., Longo, E., Leite, E.R.: The role of oriented attachment mechanism on phase transformation in oxide nanocrystals. Chem. Eur. J. **13**, 5798–5803 (2007)
22. Mullins, W.W.: 2-dimensional motion of idealized grain boundaries. J. Appl. Phys. **27**, 900–904 (1956)
23. Feltham, P.: Grain growth in metals. Acta Metall. **5**, 97–105 (1957)
24. Hillert, M.: On theory of normal and abnormal grain growth. Acta Metall. **13**, 227 (1965)
25. Louat, N.P.: Theory of normal grain-growth. Acta Metall. **22**, 721–724 (1974)
26. Herrmann, G., Gleiter, H., Baro, G.: Investigation of low-energy grain-boundaries in metals by a sintering technique. Acta Metall. **24**, 353–359 (1976)
27. Sautter, H., Gleiter, H., Baro, G.: Effect of solute atoms on energy and structure of grain-boundaries. Acta Metall. **25**, 467–473 (1977)

28. Erb, U., Gleiter, H.: Effect of temperature on the energy and structure of grain-boundaries. Scr. Metall. **13**, 61–64 (1979)

29. Kuhn, H., Baero, G., Gleiter, H.: Energy-misorientation relationship of grain-boundaries. Acta Metall. **27**, 959–963 (1979)

30. Niederberger, M., Colfen, H.: Oriented attachment and mesocrystals: non-classical crystallization mechanisms based on nanoparticle assembly. Phys. Chem. Chem. Phys **8**, 3271–3287 (2006)

31. Dalmaschio, C.J., Ribeiro, C., Leite, E.R.: Impact of the colloidal state on the oriented attachment growth mechanism. Nanoscale **2**, 2336 (2010)

32. Leite, E.R., Giraldi, T.R., Pontes, F.M., Longo, E., Beltran, A., Andres, J.: Crystal growth in colloidal tin oxide nanocrystals induced by coalescence at room temperature. Appl. Phys. Lett. **83**, 1566–1568 (2003)

33. Zhang, J., Huang, F., Lin, Z.: Progress of nanocrystalline growth kinetics based on oriented attachment. Nanoscale **2**, 18–34 (2010)

34. Ribeiro, C., Lee, E.J.H., Longo, E., Leite, E.R.: A kinetic model to describe nanocrystal growth by oriented attachment mechanism. Chemphyschem **6**, 690–696 (2005)

35. Penn, R.L.: Kinetics of oriented aggregation. J. Phys. Chem. B **108**, 12707–12712 (2004)

36. Yang, H.G., Zeng, H.C.: Self-Construction of Hollow SnO_2 Octahedra based on two-dimensional aggregation of nanocrystallites. Angew. Chem. Int. Ed. **43**, 5930 (2004)

37. Colfen, H., Antonietti, M.: Mesocrystals and nonclassical crystallization. Wiley, Hoboken (2008)

38. Zhang, J., Lin, Z., Lan, Y.Z., Ren, G.Q., Chen, D.G., Huang, F., Hong, M.C.: A multistep oriented attachment kinetics: coarsening of ZnS nanoparticle in concentrated NaOH. J. Am. Chem. Soc. **128**, 12981–12987 (2006)

39. Stroppa, D.G., Montoro, L.A., Beltran, A., Conti, T.G., da Silva, R.O., Andres, J., Longo, E., Leite, E.R., Ramirez, A.J.: Unveiling the chemical and morphological features of Sb–SnO_2 nanocrystals by the combined use of high-resolution transmission electron microscopy and ab initio surface energy calculations. J. Am. Chem. Soc. **131**, 14544 (2009)

40. Yuwono, V.M., Burrows, N.D., Soltis, J.A., Penn, R.L.: Oriented aggregation: formation and transformation of mesocrystal intermediates revealed. J. Am. Chem. Soc. **132**, 2163–2165 (2010)

41. Tang, Z.Y., Zhang, Z.L., Wang, Y., Glotzer, S.C., Kotov, N.A.: Spontaneous self-assembly of CdTe. Nanocrystals into free-floating sheets. Science **314**, 274 (2006)

42. Godinho, M., Ribeiro, C., Longo, E., Leite, E.R.: Influence of microwave heating on the growth of gadolinium-doped cerium oxide nanorods. Cryst. Growth Des. **8**, 384–386 (2008)

43. Ribeiro, C., Lee, E.J.H., Giraldi, T.R., Varela, J.A., Longo, E., Leite, E.R.: Study of synthesis variables in the nanocrystal growth behavior of tin oxide processed by controlled hydrolysis. J. Phys. Chem. **108**, 15612–15617 (2004)

44. Weiss, G.H.: Overview of theoretical models for reaction rates. J. Stat. Phys. **42**, 3–36 (1986)

45. von Smoluchowski, M.: Versuch einer mathematischen Theorie der Koagulationkinetik kollider lösungen. Z. Phys. Chem. Stoechiom. Verwandtschaftsl. **29**, 129–168 (1917)

46. Mozyrsky, D., Privman, V.: Diffusional growth of colloids. J. Chem. Phys. **110**, 9254–9258 (1999)

47. Huang, F., Zhang, H.Z., Banfield, J.F.: The role of oriented attachment crystal growth in hydrothermal coarsening of nanocrystalline ZnS. J. Phys. Chem. B **107**, 10470–10475 (2003)

48. Huang, F., Zhang, H.Z., Banfield, J.F.: Two-stage crystal-growth kinetics observed during hydrothermal coarsening of nanocrystalline ZnS. Nano Lett. **3**, 373–378 (2003)

49. Huang, F., Gilbert, B., Zhang, H., Finnegan, M.P., Rustad, J.R., Kim, C.S., Waychunas, G.A., Banfield, J.F.: Interface interactions in nanoparticle aggregates. Geochim. Cosmochim. Acta **68**, A222–A222 (2004)

50. Drews, T.O., Katsoulakis, M.A., Tsapatsis, M.: A mathematical model for crystal growth by aggregation of precursor metastable nanoparticles. J. Phys. Chem. B **109**, 23879–23887 (2005)

51. Drews, T.O., Tsapatsis, M.: Model of the evolution of nanoparticles to crystals via an aggregative growth mechanism. Microporous Mesoporous Mater. **101**, 97–107 (2007). International Symposium on Zeolite and Microporous Crystals 2006

52. Zhang, H.Z., Huang, F., Gilbert, B., Banfield, J.F.: Molecular dynamics simulations, thermodynamic analysis, and experimental study of phase stability of zinc sulfide nanoparticles. J. Phys. Chem. B **107**, 13051–13060 (2003)

53. Lee, E.: Síntese e Caracterização de Nanopartículas de Óxido de Estanho obtidas a partir de Suspensões Coloidais. Master's thesis, University Federal of São Carlos (2004)

54. Zhang, H.Z., Banfield, J.F.: Kinetics of crystallization and crystal growth of nanocrystalline anatase in nanometer-sized amorphous titania. Chem. Mater. **14**, 4145–4154 (2002)

55. Leite, E.R., Vila, C., Bettini, J., Longo, E.: Synthesis of niobia nanocrystals with controlled morphology. J. Phys. Chem. B **110**, 18088–18090 (2006)

56. Kukushkin, S.A., Osipov, A.V.: Kinetics of first-order phase transitions in the asymptotic stage. J. Exp. Theor. Phys. **86**(6), 1201–1208 (1998)

57. Pesika, N.S., Hu, Z.S., Stebe, K.J., Searson, P.C.: Quenching of growth of ZnO nanoparticles by adsorption of octanethiol. J. Phys. Chem. B **106**, 6985–6990 (2002)

58. Oskam, G., Hu, Z.S., Penn, R.L., Pesika, N., Searson, P.C.: Coarsening of metal oxide nanoparticles. Phys. Rev. E: Stat. Nonlinear Soft Matter Phys. **66**, 011403-1-4 (2002)

59. Hu, Z.S., Oskam, G., Penn, R.L., Pesika, N., Searson, P.C.: The influence of anion on the coarsening kinetics of ZnO nanoparticles. J. Phys. Chem. B **107**, 3124–3130 (2003)

60. Oskam, G., Nellore, A., Penn, R.L., Searson, P.C.: The growth kinetics of TiO$_2$ nanoparticles from titanium(IV) alkoxide at high water/titanium ratio. J. Phys. Chem. B **107**, 1734–1738 (2003)

61. Hu, Z.S., Oskam, G., Searson, P.C.: Influence of solvent on the growth of ZnO nanoparticles. J. Colloid Interface Sci. **263**, 454–460 (2003)

62. Hu, Z.S., Ramirez, D.J.E., Cervera, B.E.H., Oskam, G., Searson, P.C.: Synthesis of ZnO nanoparticles in 2-propanol by reaction with water. J. Phys. Chem. B **109**, 11209–11214 (2005)

63. Hu, Z.S., Santos, J.F.H., Oskam, G., Searson, P.C.: Influence of the reactant concentrations on the synthesis of ZnO nanoparticles. J. Colloid Interface Sci. **288**, 313–316 (2005)

64. Ribeiro, C., Lee, E.J.H., Longo, E., Leite, E.R.: Oriented attachment in anisotropic nanocrystals: a "polimeric" approach. Chemphyschem **7**, 664–670 (2006)

65. Chalmers, W.: The mechanism of macropolymerization reactions. J. Am. Chem. Soc. **56**, 912–922 (1934)

66. Flory, P.J.: Molecular size distribution in linear condensation polymers. J. Am. Chem. Soc. **58**, 1877–1885 (1936)

67. Kuchanov, S., Slot, H., Stroeks, A.: Development of a quantitative theory of polycondensation. Prog. Polym. Sci. **29**, 563–633 (2004)

68. Jun, Y.W., Casula, M.F., Sim, J.H., Kim, S.Y., Cheon, J., Alivisatos, A.P.: Surfactant-assisted elimination of a high energy facet as a means of controlling the shapes of TiO$_2$ nanocrystals. J. Am. Chem. Soc. **125**, 15981–15985 (2003)

69. Cozzoli, P., Kornowski, A., Weller, H.: Low-temperature synthesis of soluble and processable organic-capped anatase TiO$_2$ nanorods. J. Am. Chem. Soc. **125**, 14539–14548 (2003)

70. Cho, K.S., Talapin, D.V., Gaschler, W., Murray, C.B.: Designing PbSe nanowires and nanorings through oriented attachment of nanoparticles. J. Am. Chem. Soc **127**, 7140–7147 (2005)

71. Liu, K., Nie, Z., Zhao, N., Li, W., Rubinstein, M., Kumacheva, E.: Step-growth polymerization of inorganic nanoparticles. Science **329**, 197–200 (2010)

72. Niederberger, M., Pinna, N.: Metal Oxide Nanoparticles in Organic Solvents – Synthesis, Formation, Assembly and Applications. Springer, London (2009)

73. Garnweitner, G., Goldenberg, L., Sakhno, O., Antonietti, M., Niederberger, M., Stumpe, J.: Large-scale synthesis of organophilic zirconia nanoparticles and their application in organic-inorganic nanocomposites for efficient volume holography. Small **3**, 1626–1632 (2007)

Chapter 5
Oriented Attachment (OA) with Solid–Solid Interface

One of the characteristics of the OA mechanism not found in the OR (Ostwald ripening) mechanism is the presence of a solid–solid interface between nanocrystals, indicating that the growth process begins only after contact is established between particles. Two types of interface can be generated by collision:

(a) Type I – This interface is characterized by a coherent crystallographic orientation between the particles in contact.
(b) Type II – In this type of interface, the particles have no common crystallographic orientation.

In the type II interface, the prerequisite that the primary particle be in compatible orientation does not apply and, to achieve structural coherence at the interface, the primary particle must rotate into a common crystallographic orientation. This process of attachment is followed by a rotation process, which leads to a low-energy configuration, thereby forming a coherent grain boundary. After the rotation, the type II interface transforms into a type I interface. When this rotation process takes place, the growth process is also known as the grain-rotation-induced grain coalescence (GRIGC) mechanism.

A step common to all the types of OA mechanisms described so far is the elimination of grain boundaries after the formation of a type I interface. Elimination of the grain boundary produces a single larger nanocrystal, and this step is called the self-recrystallization process. The next section describes a quantitative analysis of the GRIGC mechanism and the different processes that occur after the formation of the solid-solid interface by the process of OA.

5.1 Quantitative Analysis of the GRIGC Mechanism

In a recent and remarkable in situ TEM study, Zheng et al. [1] followed the nucleation and growth of Pt nanocrystals by reducing the Pt cations with electron beams. Their recorded images allow for a statistical analysis (frame to frame) of

E.R. Leite and C. Ribeiro, *Crystallization and Growth of Colloidal Nanocrystals*, 69
SpringerBriefs in Materials, DOI 10.1007/978-1-4614-1308-0_5,
© Edson Roberto Leite and Caue Ribeiro 2012

Fig. 5.1 Schematic representation of nonepitaxial attachment between nanoparticles, followed by particle rotation and alignment

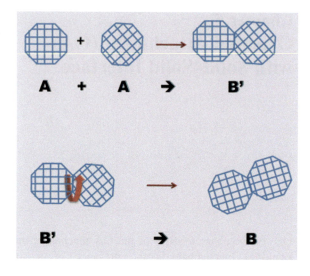

the growth trajectories of nanocrystals. The authors found that each nanocrystal has a finite probability of either growing through the addition of monomers from solution or merging with another nanocrystal in random coalescence (epitaxial and nonepitaxial growth) events and thereby jumping ahead in the growth race. Epitaxial growth can be understood as an oriented attachment process. Perhaps Zheng et al.'s most important observation is that coalescence events are often nonepitaxial (which would create a single-crystalline particle), resulting in a solid–solid interface between nanoparticles. After the imperfect attachment of the nanocrystals, the composite particle appears to undergo reorganization, i.e., particle rotation and alignment. Figure 5.1 schematically represents nonepitaxial attachment between nanocrystals, followed by particle rotation and alignment. This result strongly suggests that particle rotation and alignment is an important step in the growth mechanism during the synthesis of nanocrystals.

Several other reports of in situ and ex situ experiments have shown that particle growth can occur by means of relative rotations between the particles or by plastic deformation associated with displacement motion, until a thermodynamically favorable interface configuration (i.e., crystallographic alignment) is reached. The process was modeled by Moldovan et al. [2–5], investigated by molecular dynamics studies [6–9], and confirmed experimentally [10–13]. In all these theoretical studies, the authors assumed that the nanoparticles are in contact with each other, i.e., there is a well-defined solid–solid interface.

The model assumes the existence of a *cumulative torque*, i.e., the total torque energy of the particles in contact, given by the variation in surface energy. Accordingly, the lowest total surface energy in a system would be that at which all the particles are aligned crystallographically. Thus, the cumulative torque τ_{cum} at the center of mass of the particles is [4, 12]:

$$\tau_{\text{cum}} = \Sigma_i L_i \mathrm{d}\gamma_i / \mathrm{d}\theta_i, \tag{5.1}$$

where L_i is the length of each individual boundary i with surface energy γ_i, and θ_i is the mismatch angle between a particle and its neighbor i. Since the process occurs on surfaces in contact, the interaction between particles is the mobility of crystalline defects (surface discordances). This is a viscous process, which can be defined as inversely proportional to the particle radius, with an adjusting exponent. Thus, the angular velocity ω of a single particle is given by:

$$\omega \propto (1/r^g)\tau_{cum}, \tag{5.2}$$

where g is an adjustment exponent. When two neighboring particles rotate in the same crystallographic orientation, they coalesce, forming a new particle, and the total number of particles N_p in the system will decay discontinuously in each coalescence event. Thus, assuming a characteristic lifetime for the process, t_L, equal to the average time for a coalescence event, one has:

$$(1/N_p)dN_p/dt = -1/t_L. \tag{5.3}$$

A reasonable supposition is that coalescence occurs at a frequency proportional to the rotation rate, i.e., $t_L \propto <\omega>$. Observing only the two-dimensional case, assuming N_p particles confined in a square box of L_b edges, the average area of each particle A_p (transverse section) is given by $A_p \approx L_b^2/N_p$. Rearranging and differentiating, one has $N_p dA_p = -A_p dN_p$, or $dA_p/A_p = -dN_p/N_p$. Thus, Eq. 5.3 is rewritten as a function of the transverse section of each particle or as a function of A_p:

$$(1/A_p)dA_p/dt \propto <\omega>. \tag{5.4}$$

The value $<\omega>$ is defined by Eq. 5.2; for the sake of simplicity, the cumulative torque is rewritten as $\tau_{cum} \approx A_p^{1/2}\gamma'$, where γ' is the average value of $d\gamma_i/d\theta_i$, as defined in Eq. 5.1. Then, substituting the approximations and integrating, one has:

$$A_p \propto t^{2/(g-1)}. \tag{5.5}$$

The authors made a molecular dynamics study of the exponent of g which they adopted for two boundary conditions: $g = 4$ is related to diffusion inside particles, and $g = 5$ is related to the rotation process [6–8]. Using the value $g = 5$, the equation is simplified to $A_p \propto t^{1/4}$ [4, 9]. This equation may be applied to explain the columnar growth of nanostructures in a single direction typically occurring in thin films. In this case, Eq. 5.5 may be rewritten according to the length of the bottom particle to obtain a relation with the particle height $L_g \propto t^{2/(g-1)} = t^{1/2}$.

Using a thermodynamic approach, a very similar analysis was developed by Thompson [14–17] and adapted by Leite et al. [18]. The model was proposed to explain secondary grain growth (abnormal growth) in thin films. In this model, a single crystal with flat surfaces is bound to a matrix comprised of an array of crystals, as shown in Fig. 5.2. The growth process leads to the formation of an

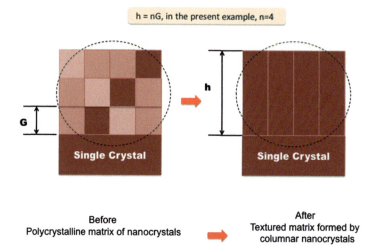

Fig. 5.2 Schematic of the cross-sectional view of a crystal array before and after growth

oriented film with thickness h. Consider a cylindrical grain (assumed here as polycrystalline) of radius r_s and thickness h growing into a uniform matrix characterized by a grain boundary energy per unit volume of E_b given by:

$$E_b = N \cdot A_b \cdot \gamma_b, \tag{5.6}$$

where N is the total number of particles per unit volume, A_b is the total particle boundary area, and γ_b is the mean boundary energy per unit area. Considering a region with a bottom area πr^2 and height h, as shown in Fig. 5.2 before and after the growth process, the energy per unit volume is expressed by:

$$F_i = (2\pi \cdot r_s^2) \cdot \gamma + (\pi r_s^2 h) N \cdot A_b \cdot \gamma_b / \pi r_s^2 h, \tag{5.7}$$

$$F_f = (2\pi \cdot r_s^2) \cdot \gamma_{min} + (2\pi \cdot r_s \cdot h) \cdot \gamma_b / \pi r_s^2 h. \tag{5.8}$$

The energy variation during the growth process, $\Delta F = F_f - F_i$, is the driving force per unit volume for crystal growth and is given by:

$$\Delta F = -2\Delta\gamma / h - N \cdot A_b \cdot \gamma_b + 2\gamma_b / r_s \tag{5.9}$$

where $\Delta\gamma = \gamma_a - \gamma_{min}$ is the surface energy anisotropy, or the variation of the surface energy per unit area (γ_a is the mean surface energy per unit area and γ_{min} the surface energy of the oriented crystal), and h is the height of the thin film. The first term (Eq. 5.9) relates to the crystallographic anisotropy of the single crystal; the second, to the growth process; and the third represents a barrier to the growth process, afforded by secondary grain boundary energy.

To simplify the expression, one can suppose that the matrix is composed of an array of uniform hexagonal grains with dimensions r_n. Substituting N, A_b for the expressions relating to the geometrical shape, one has:

$$N = 2/[3\sqrt{(3)} \cdot r_n^2 \cdot h], \tag{5.10}$$

$$A_b = 3 \cdot r_n h. \tag{5.11}$$

Substituting the values in Eq. 5.9, one has:

$$\Delta F = -2\Delta g/h - b\gamma_b/r_n + 2\gamma_b/r_n, \tag{5.12}$$

where $b = 1.15$. Leite and coauthors [18] stated that an oriented thin film growth process may be given by:

$$dh/dt \propto M_b \cdot \Delta F, \tag{5.13}$$

where M_b is boundary mobility. During the growth process, one can consider that a relation exists between mean nanocrystal size (G) and h. It is postulated that h scales linearly with crystal area, i.e.:

$$h \propto n \cdot r_n^2, \tag{5.14}$$

where $n = 1, 2, 3....$. In this model, it is assumed that the growth process occurs stepwise and is controlled by a coalescence mechanism. Each step is proportional to the mean nanocrystal size. Substituting the equations above in Eq. 5.12, one has:

$$dh/dt \propto n \cdot M_b/h - (nb - 1)\gamma_b - \Delta\gamma. \tag{5.15}$$

In Eq. 5.15, one can see that the growth rate is dependent on an integer number of grains, n. The equation can be integrated to obtain:

$$h \approx k \cdot t^{1/2}, \tag{5.16}$$

where k is a constant. This result is very similar to that obtained by Moldovan et al. [2–5] assuming grain boundary diffusion as the accommodation mechanism for particle rotation.

Several studies in the literature suggest that the OA process may occur during the classical crystallization process from an amorphous inorganic phase, as reported by Leite and coauthors for the crystallization and growth mechanism of Nb_2O_5 (niobia) treated at low temperature [19]. In his work, HRTEM characterization showed the presence of well-crystallized nanorods of TT-niobia phase, with preferential growth in the [001] direction, originated from an amorphous inorganic precursor. Another example is the crystallization of $CaCO_3$ [20]. Using cryo-TEM,

Sommerdijk and coauthors found that template-directed $CaCO_3$ synthesis starts with the formation of prenucleation clusters. The aggregation of these clusters leads to the formation of amorphous nanoparticles in solution. During the crystallization process of the amorphous nanoparticles, they observed the formation of a crystalline domain stabilized by the template. After this step, they reported the formation and growth of oriented single crystals, suggesting the occurrence of OA during the growth process.

5.2   The Self-recrystallization Process

"Recrystallization," which is a general term used to describe a solid–solid transformation, can be classified into two types:

(a) Primary recrystallization – In this type of solid phase transformation process, new crystallites are nucleated and then grow at the expense of the deformed structure until the imperfect material is consumed.
(b) Secondary recrystallization – In this case, grain boundary migration is restricted to a minority of boundaries. Hence, only a few crystallites grow at the expense of all the rest.

In the OA mechanism, the self-recrystallization phenomenon can be classified as a secondary recrystallization process dominated by grain boundary migration. In fact, this is a special type of secondary recrystallization process in which all the particles or grains involved in the process share a common crystallographic orientation.

It is well documented in the literature that after the recrystallization step, the material grown by OA presents a defined shape [21–23]. For instance, Giersig and coauthors [22] described the spontaneous organization of single CdTe nanocrystals into a pearl-like one-dimensional structure and the transformation of the pearl-like structures into nanowires with a single crystal structure and a well-defined shape. Very recently, Penn and coauthors [23] reported the formation of mesocrystals composed of oriented goethite nanocrystals as a necessary precursor to the formation of single-crystal oriented aggregates of goethite. After self-recrystallization, a well-defined nanorod with single-crystal characteristics was reported. Thus, two events occur during the self-recrystallization process, namely, the elimination of grain boundaries (resulting in a material with a single-crystalline domain) and the development of the final shape of the crystals. Figure 5.3 represents the events involved in the self-recrystallization process. The HRTEM images of CeO_2 nanowires before and after recrystallization illustrate the elimination of the grain boundary and the modification of the wire shape, with the formation of a smooth surface. These images clearly illustrate the effect of grain boundary elimination on the particle shape. Since this shape is defined by elimination of the interface, it can be assumed that the final shape of the nanocrystal is dictated by the minimization of surface energy.

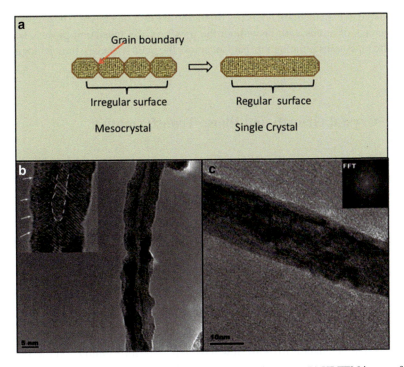

Fig. 5.3 (**a**) Schematic representation of the recrystallization process. (**b**) HRTEM image of the CeO₂ nanowire before recrystallization, showing the presence of grain boundaries and irregular surfaces. The inset depicts the reconstructed lattice image of the wire, showing the boundaries among oriented nanoparticles. (**c**) HRTEM image of the CeO₂ nanowire after recrystallization. The inset shows an FFT image typical of a single crystal material

A typical characteristic of the recrystallization process is that it hinders any movement (migration) of atoms or nanocrystals in the phase. Therefore, the elastic deformation that develops during the transformation process should be taken into account. In the self-recrystallization process reported here, a mesocrystal is transformed into a single crystal. The change in free energy (ΔG) at the transformation of the new single crystal can be described as:

$$\Delta G = -\Delta G'v + \sum \gamma_{ss} S + \Phi_{def}, \tag{5.17}$$

where $\Delta G'$ is the change in the Gibbs free energy G' of the initial state (mesocrystal) and final (single crystal) transformation, v is the volume, γ_{ss} and S are the specific surface energy and areas of interfaces (grain boundaries), respectively, and Φ_{def} is the term of the elastic deformation energy. It is proposed that the second and elastic energy terms determine the crystal's final morphology and that the shape dictated by minimization of surface energy is controlled mainly by the change in interface energy (the second term of Eq. 5.17). As a result of the limited mobility of the atoms or nanocrystals in the solid phase, cooperative movements of atoms should be considered to explain grain boundary migration at low temperature. These cooperative movements of atoms with or without the presence of a relaxation process lead to a specific mass transportation mechanism across the grain boundary (not yet identified).

5.3   Crystal Growth and Phase Transformation

The correlation between crystalline phase and size was widely explored in different studies, which propose that many differences in terms of phase stability may be explained by crystal size. In fact, it is generally understood that many metastable structures in nanomaterials appear to be stable in the nanometric range without the addition of dopants or constriction by a matrix [24]. A typical case is the synthesis of nanocrystalline zirconium oxide in the tetragonal phase. Although tetragonal zirconia appears stable only over 1,175°C, early reports attested to the stabilization of the phase below 10 nm without applied tension, and stabilization below 40 nm when constricted by a matrix [25–28]. In fact, few works have reported monoclinic zirconia in the nanometric range, except in the 15-nm range [29].

Another case commonly reported is the synthesis of TiO_2 polymorphs, anatase, brookite, and rutile [30]. While several papers report on the synthesis of nanocrystalline anatase [31–36], few have reported, for example, on the synthesis of nanocrystalline rutile [37, 38]. However, several papers state that the formation of rutile passes through the three metastable phases, and it has been established that rutile is the most stable TiO_2 polymorph (observations of micrometric anatase are scarce) [24, 39–42]. This sequence (from the most unstable to the stable polymorph) is called the Ostwald step rule, and it is widely accepted as a general rule for crystal formation.

Those typical cases are generally explained in terms of the thermodynamic properties of solid phases. Many authors argue that the surface energy of metastable phases is always lower than that of the most stable polymorph. In the case of ZrO_2 stability, Pitcher and coauthors [43] discussed in detail the transition of tetragonal to monoclinic based on calorimetric surface energy enthalpy measurements, demonstrating that the average surface energy of the tetragonal phase is lower than that of the monoclinic phase. Thus, since the classical nucleation theory postulates that the stability of a given nucleus or small cluster is given by the balance between the free energy of formation (favorable to the nucleus formation) and the work given by the new surface (unfavorable), it is clear that lower surface energies will favor the formation of these metastable crystals in smaller sizes. This simple formulation can explain, with good results, some of the observed cases. Garvie [25] used this argument to explain the tetragonal stabilization of nanocrystals at room temperature, developing a thermodynamic approach to determine a critical size for phase stabilization.

TiO_2 can be analyzed similarly. In principle, anatase is metastable in relation to rutile at any temperature [30, 44], although the energetics of anatase to rutile transformation can be estimated for a given temperature. Gribb and Banfield [45] observed experimentally that the critical size for the stability of anatase crystals was 14 nm, but phase transformation occurred at larger sizes and the resulting rutile nanocrystals grew relatively faster. In fact, anatase nanocrystals are commonly observed in sizes below 15 nm [45–48], although several sizes from 15 to 50 nm [46–51] have also been reported in the literature. This is explained by the dependence of anatase stabilization of several parameters, such as the starting material and the environment and temperature of synthesis.

An interesting generalization of the discussion is to assume that phase stability depends on global thermodynamic parameters, i.e., the total surface energy of the particle is a function of its volume. Barnard and Zapol [52, 53] and Barnard et al. [54, 55] engaged in similar discussions, obtaining a general expression for the free energy, G_x^o, of an arbitrary nanoparticle, taking into account the balance of surface energy as a function of the exposed crystallographic planes, as follows:

$$G_x^0 = G_x^{\text{bulk}} + M/\rho_x(1 - e)q \sum f_i \gamma_i(T), \tag{5.18}$$

where the first term is defined as the standard free energy of formation, $G_x^{\text{bulk}}(T)$, and the second term is expressed in terms of surface energy γ_i for each i plane on the surface and molar surface area A. This can be described using the relations of density of phase $\times$ (ρ_x), molar mass M, the volume dilation of the nanoparticle e (negligible in several cases), the surface to volume ratio q, and f_i, a weight factor of the facets i in the crystal $(\Sigma f_i = 1)$. In the above formulation, the expression takes into account the crystallographic alignment of the properties and, indirectly, the shape. In this formulation, one can add the contributions of factors such as the interaction with ions on the surface (counterions, surfactants, etc.) as a way to minimize energy in specific crystallographic planes.

It is important to emphasize that in this formulation, the shape – which defines the number of facets, edges, and corners – is the determining factor of phase stability. Using quantum mechanical calculations, Barnard et al. confirmed this assumption by demonstrating several stable shapes for tetragonal ZrO_2 nanocrystals with different limiting sizes. Also, it is worth noting that although several papers report long anatase nanorods, anatase nanospheres larger than 14 nm are scarce – note the influence of shape in both cases, which explains retention of the metastable phase in this case.

Since anisotropy is strongly dependent on the growth mechanisms involved in nanocrystal formation, it can play an important role in phase transformation. Many studies attribute the growth of isotropic nanocrystals to the Ostwald ripening mechanism [56–60]. In this mechanism, the particles tend toward an isotropic growth due to their original atomic arrangement, generating particles with regular quasi-spherical shapes. Hence, variations in the weight factor, f_i, of the facets are not expected. However, crystal growth by oriented attachment [61–63] has been demonstrated to favor the formation of anisotropic nanocrystals through the coalescence of two or more nanocrystals. Previous theoretical studies [64–66] emphasize that this mechanism is faster than the diffusional mechanism, particularly in the early stages of growth. Thus, one can assume that the rapid growth and anisotropic shapes obtained may give rise to other shape and size-related phenomena such as spontaneous phase transformation.

Ribeiro and coauthors [67, 68] investigated this possible correlation in different phases of titanium and zirconium oxide nanocrystals by means of HRTEM. To this end, titanium and zirconium oxides were synthesized by hydrothermally treating a gel solution of peroxo complexes of titanium (PCT) and zirconium (PCZ). The authors obtained TiO_2 rutile and anatase of comparable sizes, and tetragonal and monoclinic ZrO_2 in the same conditions. They found that stabilization of the most stable polymorph in sizes below the critical was only possible when highly anisotropic structures (such as rods or wires) were obtained. Therefore, they concluded that phase stability was strongly dependent on minimization of the total surface energy. Tailoring highly anisotropic crystals, favoring planes with low energy, could attain this minimization. The OA mechanism can interfere by forming anisotropic particles, particularly when growth occurs preferentially in the higher energy planes.

The role of oriented attachment in phase stability is to stabilize metastable phases. This role probably relies on methods to prevent growth by oriented attachment, whereby the stabilization of stable phases in nanometric range can be accelerated by inducing oriented attachment on the particles' high energy planes.

From this discussion, the role of the growth mechanism in phase control can be understood in terms of its influence on the weight factor, f_i, of the facets, which can be attained by tailoring anisotropic structures (since the OA mechanism is related to this aspect). Its influence is interpreted as the modification of the area to volume ratio in newly formed particles, favoring phase transformation, or not, according to the crystallographic planes exposed after the event. However, the surface energy of each plane is strongly affected by the presence of counterions in the medium. In the synthesis of TiO_2, it has been shown that the presence of common ions in the synthesis environment, such as Cl^- or organic chains from precursors, can alter phase stability [58].

References

1. Zheng, H., Smith, R.K., Jun, Y.-W., Kisielowski, C., Dahmen, U., Alivisatos, A.P.: Observation of single colloidal platinum nanocrystal growth trajectories. Science **5**, 1309 (2009)
2. Moldovan, D., Wolf, D., Phillpot, S.R.: Theory of diffusion-accommodated grain rotation in columnar polycrystalline microstructures. Acta Mater. **49**, 3521 (2001)
3. Moldovan, D., Wolf, D., Phillpot, S.R., Haslam, A.J.: Role of grain rotation during grain growth in a columnar microstructure by mesoscale simulation. Acta Mater. **50**, 3397 (2002)
4. Moldovan, D., Yamakov, V., Wolf, D., Phillpot, S.R.: Scaling behavior of grain-rotation-induced grain growth. Phys Rev Lett **89**, 206101 (2002)
5. Haslam, A.J., Moldovan, D., Yamakov, V., Wolf, D., Phillpot, S.R., Gleiter, H.: Stress-enhanced grain growth in a nanocrystalline material by molecular-dynamics simulation. Acta Mater. **51**, 2097 (2003)

6. Jensen, P.: Growth of nanostructures by cluster deposition: Experiments and simple models. Rev. Mod. Phys. **71**, 1695 (1999)

7. Zhu, H.L., Averback, R.S.: Sintering of nano-particle powders: Simulations and experiments. mater. Manuf. Processes **11**, 905 (1996)

8. Zhu, H.L., Averback, R.S.: Sintering processes of two nanoparticles: A study by molecular dynamics simulations. Philos. Mag. Lett. **73**, 27 (1996)

9. Zhang, H.Z., Huang, F., Gilbert, B., Banfield, J.F.: Molecular dynamics simulations, thermodynamic analysis, and experimental study of phase stability of zinc sulfide nanoparticles. J Phys Chem B **107**, 13051 (2003)

10. Yeadon, M., Yang, J.C., Averback, R.S., Bullard, J.W., Olynick, D.L., Gibson, J.M.: In-situ observations of classical grain growth mechanisms during sintering of copper nanoparticles on (001) copper. Appl. Phys. Lett. **71**, 1631 (1997)

11. Yeadon, M., Ghaly, M., Yang, J.C., Averback, R.S., Gibson, J.M.: "Contact epitaxy" observed in supported nanoparticles. Appl. Phys. Lett. **73**, 3208 (1998)

12. Harris, K.E., Singh, V.V., King, A.H.: Grain rotation in thin films of gold. Acta Mater. **46**, 2623 (1998)

13. Ribeiro, C., Lee, E.J.H., Giraldi, T.R., Aguiar, R., Longo, E., Leite, E.R.: *In situ* oriented crystal growth in a ceramic nanostructured system. J. Appl. Phys. **97**, 024313 (2005)

14. Thompson, C.V.: Secondary grain growth in thin films of semiconductors: Theoretical aspects. J. Appl. Phys. **58**, 763 (1985)

15. Thompson, C.V., Carel, R.: Texture development in polycrystalline thin films. Materials Science and Engineering: B-Advanced Functional Solid-State Materials **32**, 211 (1995)

16. Thompson, C.V., Smith, H.I.: Surface-energy-driven secondary grain growth in ultrathin (<100 nm) films of silicon. Appl. Phys. Lett. **44**, 603 (1984)

17. Wong, C.C., Smith, H.I., Thompson, C.V.: Surface-energy-driven secondary grain growth in thin Au films. Appl. Phys. Lett. **48**, 335 (1986)

18. Leite, E., Khan, A., Scotch, A.M., Chan, H.M., Harmer, M.P.: Proceeding of the sintering Conference 1999, 355, Penn State University, State College PA, (2000)

19. Leite, E.R., Vila, C., Bettini, J., Longo, E.: Synthesis of niobia nanocrystals with controlled morphology. J Phys Chem B **110**, 18088 (2006)

20. Pouget, E.M., Bomans, P.H.H., Goos, J.A.C.M., Frederik, P.M., de With, G., Sommerdijk, N.A.J.M.: The initial stages of template-controlled $CaCO_3$ formation revealed by cryo-TEM. Science **13**, 1455 (2009)

21. Burrows, N.D., Yuwono, V.M., Penn, R.L.: Quantifying the kinetics of crystal growth by oriented aggregation. MRS Bull. **35**, 133 (2010)

22. Tang, Z., Kotov, N.A., Giersig, M.: Spontaneous organization of single cdte nanoparticles into luminescent nanowires. Science **12**, 237 (2002)

23. Yuwono, V.M., Burrows, N.D., Soltis, J.A., Penn, R.L.: Oriented aggregation: Formation and transformation of mesocrystal intermediates revealed. J Am Chem Soc **132**, 2163 (2010)

24. Navrotski, A.: Energetic clues to pathways to biomineralization: Precursors, clusters, and nanoparticles. Proc Natl Acad Sci USA **101**, 12096 (2004)

25. Garvie, R.C.: The occurrence of metastable tetragonal zirconia as a crystallite size effect. J Phys Chem **69**, 1238 (1965)

26. Garvie, R.C.: Phase analysis in zirconia systems. J. Am. Cer. Soc. **55**, 303 (1972)

27. Garvie, R.C.: Stabilization of the tetragonal structure in zirconia microcrystals. J Phys Chem **82**, 218 (1978)

28. Shukla, S., Seal, S.: Thermodynamic tetragonal phase stability in Sol–Gel derived nanodomains of pure zirconia. J Phys Chem B **108**, 3395 (2004)

29. Guo, G.Y., Chen, L.: A nearly pure monoclinic nanocrystalline zirconia. J. Solid State Chem. **178**, 1675 (2005)

30. Navrotski, A., Kleppa, O.J.: Enthalpy of the anatase-rutile transformation. J. Am. Cer. Soc. **50**, 626 (1967)

31. Barringer, E.A., Bowen, H.K.: High-purity, monodisperse TiO_2 powders by hydrolysis of titanium tetraethoxide. 1. Synthesis and physical properties. Langmuir **1**, 414 (1985)

32. Jean, J.H., Ring, T.A.: Nucleation and growth of monosized titania powders from alcohol solution. Langmuir **2**, 251 (1986)
33. Mates, T.E., Ring, T.A.: Steric stability of alkoxy-precipitated TiO_2 in alcohol solutions. Colloids Surf. **24**, 299 (1987)
34. Kavan, L., Kratochvilova, K., Gratzel, M.: Study of nanocrystalline TiO_2 (anatase) electrode in the accumulation regime. J. Electroanal. Chem. **394**, 93 (1995)
35. Garnweitner, G., Antonietti, M., Niederberger, M.: Nonaqueous synthesis of crystalline anatase nanoparticles in simple ketones and aldehydes as oxygen-supplying agents. Chem. Comm. **3**, 397 (2005)
36. Trentler, T.J., Dentler, T.E., Bertone, J.F., Agrawal, A., Colvin, V.L.: Synthesis of TiO_2 nanocrystals by nonhydrolytic solution-based reactions. J. Am. Chem. Soc. **121**, 1613 (1999)
37. Ragai, J., Lotfi, W.: Effect of preparative pH and ageing media on the crystallographic transformation of amorphous TiO_2 to anatase and rutile. Colloids Surf. **61**, 97 (1991)
38. Han, S., Choi, S.H., Kim, S.S., Cho, M., Jang, B., Kim, D.Y., Yoon, J., Hyeon, T.: Low-temperature synthesis of highly crystalline TiO_2 nanocrystals and their application to photocatalysis. Small **1**, 812 (2005)
39. Zhang, H.Z., Banfield, J.F.: Kinetics of crystallization and crystal growth of nanocrystalline anatase in nanometer-sized amorphous titania. Chem. Mater. **14**, 4145 (2002)
40. Ranade, M.R., Navrotsky, A., Zhang, H.Z., Banfield, J.F., Elder, S.H., Zaban, A., Borse, P.H., Kalkarni, S.K., Doran, G.S., Whit, J.: Colloquium paper: Nanoscience: Underlying physical Concepts and phenomena: Energetics of nanocrystalline TiO_2. Proc Natl Acad Sci USA **99**, 6476 (2002)
41. Barnard, A.S., Zapol, P.: Predicting the energetics, phase stability, and morphology evolution of faceted and spherical anatase nanocrystals. J Phys Chem B **108**, 18435 (2004)
42. Barnard, A.S., Zapol, P.: Effects of particle morphology and surface hydrogenation on the phase stability of TiO_2. Phys. Rev. B **70**, 235403 (2004)
43. Pitcher, M.W., Ushakov, S.V., Navrotsky, A., Wood, B.F., Li, G., Boerio-Goates, J., Tissue, B.M.: Energy crossovers in nanocrystalline zirconia. J. Am. Cer. Soc. **88**, 160 (2005)
44. Shannon, R.D., Pask, J.A.: Kinetics of the anatase-rutile transformation. J. Am. Cer. Soc. **48**, 391 (1965)
45. Gribb, A.A., Banfield, J.F.: Particle size effects on transformation kinetics and phase stability in nanocrystalline TiO_2. Am. Mineral. **82**, 717 (1997)
46. Yanagisawa, K., Ovenstone, J.: Crystallization of anatase from amorphous titania using the hydrothermal technique: effects of starting material and temperature. J Phys Chem B **103**, 7781 (1999)
47. Kormann, C., Bahnemann, D.W., Homann, M.R.: Preparation and characterization of quantum-size titanium dioxide. J Phys Chem **92**, 5196 (1988)
48. Penn, R.L., Oskam, G., Strathmann, T.J., Searson, P.C., Stone, A.T., Veblen, D.R.: Epitaxial assembly in aged colloids. J Phys Chem B **105**, 2177 (2001)
49. Oskam, G., Nellore, A., Penn, R.L., Searson, P.C.: The growth kinetics of TiO_2 nanoparticles from titanium(IV) alkoxide at high water/titanium ratio. J Phys Chem B **107**, 1734 (2003)
50. Li, W.J., Osora, H., Otero, L., Duncan, D.C., Fox, M.A.: Photoelectrochemistry of a substituted-$Ru(bpy)_3^{2+}$-labeled polyimide and nanocrystalline SnO_2 composite formulated as a thin-film electrode. J Phys Chem A **102**, 5333 (1998)
51. Bedja, I., Kamat, P.V., Hua, X., Lappin, A.G., Hotchandani, S.: Photosensitization of nanocrystalline ZnO films by Bis(2,2'-bipyridine)(2,2'-bipyridine-4,4'-dicarboxylic acid) ruthenium(II). Langmuir **12**, 2398 (1997)
52. Barnard, A.S., Zapol, P.: A model for the phase stability of arbitrary nanoparticles as a function of size and shape. J Chem Phys **121**, 4276 (2004)
53. Barnard, A.S., Zapol, P., Curtiss, L.A.: Modeling the morphology and phase stability of TiO_2 nanocrystals in water. J. Chem. Theory Comput. **1**, 107 (2005)
54. Barnard, A., Saponjic, Z., Tiede, D., Rajh, T., Curtiss, L.: Multi-scale modeling of titanium dioxide: Controlling shape with surface chemistry. Rev. Adv. Mater. Sci. **10**, 21 (2005)

55. Barnard, A.S., Yeredla, R.R., Xu, H.F.: Modelling the effect of particle shape on the phase stability of ZrO_2 nanoparticles. Nanotechnology **17**, 3039 (2006)
56. Feldmann, C.: Preparation of nanoscale pigment particles. Adv Mater **13**, 1301 (2001)
57. Oskam, Z.S.Hu., Penn, R.L., Pesika, N., Searson, P.C.: Coarsening of metal oxide nanoparticles. Phys. Rev. E **66**, 011403 (2002)
58. Huang, F., Zhang, H.Z., Banfield, J.F.: Two-stage crystal-growth kinetics observed during hydrothermal coarsening of nanocrystalline ZnS. Nano Lett **3**, 373 (2003)
59. Lifshitz, M., Slyozov, V.V.: The kinetics of precipitation from supersaturated solid solutions. J. Phys. Chem. Solids **19**, 35 (1961)
60. Kukushkin, S.A., Slyozov, V.V.: Crystallization of binary melts and decay of supersaturated solid solutions at the ostwald ripening stage under non-isothermal conditions. J. Phys. Chem. Solids **56**, 1259 (1995)
61. Penn, R.L., Banfield, J.F.: Morphology development and crystal growth in nanocrystalline aggregates under hydrothermal conditions: Insights from titania. Geochim. Cosmochim. Acta **63**, 1549 (1999)
62. Polleux, J., Pinna, N., Antonietti, M., Niederberger, M.: Ligand-directed assembly of preformed titania nanocrystals into highly anisotropic nanostructures. Adv Mater **16**, 436 (2004)
63. Lee, E.J.H., Ribeiro, C., Longo, E., Leite, E.R.: Oriented attachment: An effective mechanism in the formation of anisotropic nanocrystals. J Phys Chem B **109**, 20842 (2005)
64. Ribeiro, C., Lee, E.J.H., Longo, E., Leite, E.R.: A kinetic model to describe nanocrystal growth by the oriented attachment mechanism. Chemphyschem **6**, 690 (2005)
65. Drews, T.O., Katsoulakis, M.A., Tsapatsis, M.: A mathematical model for crystal growth by aggregation of precursor metastable nanoparticles. J Phys Chem B **109**, 23879 (2005)
66. Ribeiro, C., Lee, E.J.H., Longo, E., Leite, E.R.: Oriented attachment mechanism in anisotropic nanocrystals: A "polymerization" approach. Chemphyschem **7**, 664 (2006)
67. Ribeiro, C., Vila, C., Matos, J.M.E., Bettini, J., Longo, E., Leite, E.R.: Role of the oriented attachment mechanism in the phase transformation of oxide nanocrystals. Chem. Eur. J. **13**, 5798 (2007)
68. Ribeiro, C., Barrado, C.M., de Camargo, E.R., Longo, E., Leite, E.R.: Phase transformation in titania nanocrystals by the oriented attachment mechanism: The role of the pH value. Chem. Eur. J. **15**, 2217 (2009)

Chapter 6
Trends and Perspectives in Nanoparticles Synthesis

The focus of nanostructured materials is gradually shifting from the synthesis of nanocrystals with a controlled morphology and size to the organization or assembly of those nanocrystals into larger nanostructures in a natural sequence, especially in the use of nanocrystals as fundamental building blocks for the development of functional thin films and devices. In addition, the synthesis of controlled nanocrystals is still a challenge, particularly in the synthesis of transition metal oxides. In this final chapter, the trends in the synthesis of nanocrystals with controlled shapes and exposed facets will be discussed with a focus on metal oxide nanoparticles.

6.1 Trends in the Synthesis of Transition Metal Oxides

The development of new synthetic routes to obtain nanocrystals and mesocrystals with controlled shapes and reactive surfaces is of great scientific and technological interest [1–5]. Actually, the search for nanocrystals with controlled facets can be considered the main topic in nanoparticle synthesis and assembly. For instance, Shen et al. have recently demonstrated that the morphological control of Co_3O_4 was fundamental to improving the activity and stability of this oxide for CO oxidation [4]. Another good example is cerium oxide (CeO_2); i.e., the main motivation for the synthesis of ceria-based nanocrystals with a controlled morphology is the possibility of developing catalytic materials having high surface area and well-defined exposed crystal planes that exhibit optimal catalytic activity [6, 7]. For instance, the CeO_2 shape/facet termination plane was found to have a strong effect on the activity of gold/ceria catalysts developed for the water-gas shift reaction [7]. Rod-like ceria NCs with {110} and {100} exposed facet planes were identified as the most desirable morphology for gold stabilization/activation. Following the same trend, the development of other nanostructured metal oxides with controlled morphology (or facets exposed on the surface) must result in materials with superior activity and stability for several applications [2]. To illustrate the trend in nanocrystal metal oxides with

E.R. Leite and C. Ribeiro, *Crystallization and Growth of Colloidal Nanocrystals*,
SpringerBriefs in Materials, DOI 10.1007/978-1-4614-1308-0_6,
© Edson Roberto Leite and Caue Ribeiro 2012

controlled exposed facets, two important materials (TiO_2 and CeO_2) will be used as examples.

Titanium (IV) oxide (TiO_2) in the anatase phase is a key functional material with interesting sensing, photocatalytic, and photovoltaic [8–10] surface-dependent properties. The TiO_2 anatase crystal is usually dominated by {101} facets which possess the lowest surface energy (see Fig. 2.2) [11, 12]. Considering other facets, theoretical studies have demonstrated the following sequence for surface relative energies: {101} < {100} < {001}. The surface relative energy variations can basically be explained by the different chemical compositions of the facets which result in diverse degrees of broken chemical bonds on the surface.

Recently, a breakthrough in the synthesis of anatase TiO_2 crystals with {001} facets was achieved by Lu et al. [13, 14] who reported the preparation of anatase microcrystals with surfaces formed preferentially by {001} facets. On the basis of first-principle quantum chemistry calculations, the strategy used by them was the reversal of the relative stability of the facets through the use of fluoride ions during synthesis. The presence of fluoride ions favors the formation of high F-Ti bonding energy at the surface leading to a decrease in the (001) surface energy which results in more stability than the (101) surface. Then Zheng et al. used a similar approach to synthesize anatase TiO_2 nanosheets with exposed {001} facets and excellent photocatalytic performance [15]. The crystallization mechanism of the anatase with {001} exposed facets synthesized in hydrofluoric acid solution (under hydrothermal conditions) is related to monomer-by-monomer assembly; i.e., the attachment of ions/molecules to a primary nucleus. The fluorine ions must act as a selective surface poisoning agent following a classical and thus predictable crystallization process.

An alternative route to process inorganic materials is through a kinetically controlled crystallization process driven by an oriented attachment (OA) growth mechanism (see Chap. 4). In this non-classical crystallization process, the formation of a crystal controlled by monomer-by-monomer assembly is replaced by a process involving the spontaneous self-organization of adjacent nanocrystals to share a common crystallographic orientation and coalescence, i.e., by the OA growth mechanism [16, 17]. The development of synthetic routes based on a non-classical crystallization process is particularly desirable because in this approach it is not necessary to use hydrofluoric acid which is extremely corrosive and a contact poison reagent. For instance, Niederberger's group synthesized anatase TiO_2 nanowires with a diameter of 3 nm and a length of several hundred nm formed by nanocrystals assembled along the [001] direction [18, 19]. The initial anatase TiO_2 nanocrystal was synthesized in benzyl alcohol, and the OA growth along the [001] direction was achieved by the selective desorption of multidentate ligands from the {001} facets [19]. Niederberger's group used a very interesting non-hydrolytic sol-gel process based on a solvent-controlled synthesis approach using benzyl alcohol as the solvent. In the solvent-controlled synthesis approach, an organic solvent acts as the solvent and reactant which is also a control agent for particle growth and thus negates the use of surfactants [20]. In this route, the solvent provides oxygen for the formation of the metal oxide. The detailed role of the organic solvent and metal precursor is quite complex. However, in general, this synthesis approach offers the possibility for greater control of the reaction pathways on a molecular level, enabling the synthesis of nanomaterials with high crystallinity as well as well-defined and uniform particle morphologies. The organic components strongly influence the composition, size, shape, and surface properties of the inorganic product.

Leite et al. [21] recently described a kinetically controlled crystallization process assisted by an OA mechanism based on a nonaqueous sol-gel solvent-controlled synthetic method to prepare recrystallized anatase TiO_2 mesocrystals (single crystals). Specifically, these authors reacted titanium (IV) chloride ($TiCl_4$) with n-octanol to synthesize an anatase TiO_2 phase. The kinetics study revealed a multi-step and hierarchical process controlled by OA, and high-resolution transmission electron microscopy (HRTEM) analysis clearly shows that the synthesized mesocrystal has a truncated bipyramidal Wulff shape which indicates that its surface is dominated by {101} facets. The recrystallized mesocrystal had several sizes, ranging from 20 to 100 nm. The Wulff shape was developed during the recrystallization step.

The use of nanocrystals as building blocks of larger structures requires stabilized nanocrystals to prevent agglomeration and induce solubility in a suitable solvent [22]. Generally, the solubility and chemical functionality of nanocrystals is achieved by adding specific stabilizing compounds in the reaction system [22–24] or postsynthesis surface modifications using methods such as molecular exchange, amphiphilic molecules and encapsulation [25–32]. The critical issues involved in the postsynthesis surface modification approach are the time and reagent-consuming process, since it requires treatment after the nanocrystal synthesis and the selective chemical functionality induced by this treatment results in nanocrystals with selective solubility [33–35]. For example, the postsynthesis treatment that induces nanocrystal solubility in a polar solvent such as water is not suitable for promoting its solubility in a solvent with a lower solubility parameter (δ) and different degrees of hydrogen bonding strength. Leite's group developed an alternative route to process functionalized metal oxide nanocrystals in a single step process by modifying the solvent-controlled synthesis approach [36]. This process involves a single-step synthetic route to Fe_3O_4 nanocrystals which are soluble in different solvents. To synthesize grafted Fe_3O_4 nanocrystals in a single step, these researchers used a high molecular weight solvent (polyol) that is attached to the particle surface which then transfers its solubility to the particle. The polyol (a bi-functional alcohol) acts as a solvent, reactant (oxygen source), and stabilizing agent. Using polyols with different molecular weights and polarity facilitates control of the size, aggregation, solubility, and morphology of the magnetic nanocrystals. The solubility behavior of the nanocrystal is directly correlated with the adsorbed polyol in the particle surface. This polyol transfers its solubility to the nanocrystal. Figure 6.1a is a photograph of vials containing solutions of Fe_3O_4 nanocrystals synthesized in polyols with different molecular weights. PEG 8000 denotes synthesis in polyethylene glycol with a molecular weight of 8,000, and PEG 1000 was synthesized in polyethylene glycol with a molecular weight of 1,000; for T1000, the material was synthesized in Poly(1,4-butanediol) which is a polyol with lower polarity. As depicted in Fig. 6.1a, Fe_3O_4 nanocrystals synthesized in T1000 show high solubility in chloroform,

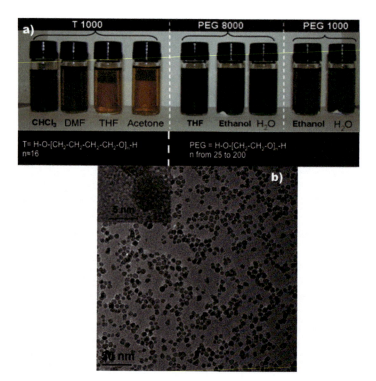

Fig. 6.1 (a) Photograph of vials containing solutions of the of a Fe_3O_4 nanocrystal synthesized in PEG 8000, PEG 1000, and T1000 in different solvents (as indicated); (**b**) low magnification HRTEM image of the material processed in PEG 1000 (inset shows a high magnification HRTEM image of the Fe_3O_4 nanocrystal)

N,N-dimethylformamide (DMF), tetrahydrofuran (THF) and acetone while the material synthesized in PEG 1000 shows solubility only in polar solvents such as water and ethanol. Leite et al. demonstrated that it is possible to modify the nanoparticle solubility by changing the polarity of the solvent used in the synthesis. The ability of the nanocrystal to become soluble in the same solvent in which T1000 shows high solubility supports the idea that the polymer layer absorbed at the particle surface is able to transfer its solubility to the nanocrystal. Actually, this result supports the argument that the conformational structure of the attached polymer is an important parameter in nanoparticle solubility. Fe_3O_4 nanocrystals synthesized in PEG 1000 and T1000 show similar morphology. Figure 6.1b depicts a low magnification HRTEM image of the material synthesized in PEG 1000. The inset shows a high magnification HRTEM image which illustrates the crystalline nature of the nanoparticle.

The modified solvent-controlled synthesis approach was also used to synthesize TiO_2 anatase nanocrystals [37] using titanium (IV) chloride ($TiCl_4$) as the metal source and poly (1,4-butanediol) with an average molecular weight of Mw = 1,000 g/mol as the solvent and oxygen source. Using this method, it was possible to obtain a

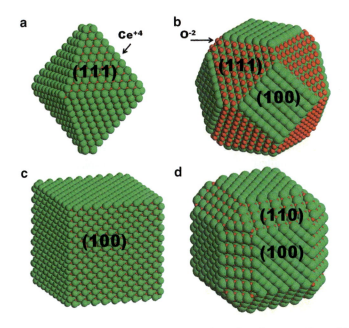

Fig. 6.2 Representation of several CeO_2 nanocrystals shapes based on crystallographic data. (**a**) Wulff reconstruction; (**b**) polyhedral; (**c**) cube; (**d**) truncated cube

TiO_2 nanocrystal with Brønsted acid sites and polymer chains chemically anchored on the nanocrystal surface. The acid surface of those nanocrystals had the chemical property to react in the presence of amine organic compounds and to maintain the colloidal stability. Actually, the titania nanoparticle surface showed the ability to transfer protons from its surface to amine compounds such as polyaniline.

The solvent-controlled synthesis approach represents an alternative and attractive route to prepare metal oxide nanocrystals with several advances over the surfactant-assistant approach and the aqueous process. For instance, in comparison to the synthesis of a metal oxide in the presence of surfactants, the solvent-controlled approach is simpler because it uses only a metal precursor and a oxygen content solvent as starting reactants and promotes crystallization at a lower temperature (typically in the range of 50–300°C) [20]. Especially in comparison with aqueous sol-gel processes, another important advantage is the possibility of synthesizing doped oxide systems or even metal oxide phases formed by two or more metals such as $SrTiO_3$, $BaTiO_3$, $(Sr,Ba)TiO_3$ [20]. Organic solvents facilitate an easier match with the reactivity of different metal precursors which is fundamental to obtaining a single-phase nanocrystal.

CeO_2 has a fluorite cubic structure, and its nanocrystals are usually dominated by {111} facets which present the lowest surface energy. Figure 6.2 depicts the representation of several CeO_2 nanocrystal shapes based on crystallographic data. Figure 6.2a illustrates Wulff construction for CeO_2 by taking into account theoretical surface energy data [38]. Figure 6.2 confirms that to promote a nanocrystal shape modification, the surface energy must be controlled. For instance, to obtain a truncated cube shape, it will be necessary decrease the surface energy of the {100} and {110} facets. A good way to promote this surface stabilization is through the use of selective adsorption of organic ligands. For example, Gao and Yang reported the synthesis of CeO_2 nanocubes with {200}-type facets using a two-phase reaction process where the size and shape of the CeO_2 nanocrystals is controlled by the amount of organic ligand or stabilizing agents (e.g., oleic acid (OAc)) was employed [39]. These authors postulate that the growth mechanism is an oriented aggregation mediated process for the larger CeO_2 nanocubes observed and a monomer deposition process for the smallest CeO_2 nanocubes. The two-phase reaction process is an interesting synthesis approach where the metal precursor is added to the polar phase (usually water) and the stabilizing agent is added to the non-polar phase (generally toluene). The nucleation process occurs in the polar phase; and with the adsorption of the ligand, the formed nanoparticles migrate to the non-polar phase. The ligand adsorption process is fundamental to control the size and shape of nanocrystals. Recently, Kuwabara et al. [40] synthesized CeO_2 nanocubes having {111} face truncations using a similar approach where they associated the growth mechanism with a difference in growth rates between the [100] and [111] directions controlled by the preferential interaction of the OAc ligands with the (100) face to limit the growth in the [100] direction which is essentially the same mechanism proposed by Gao and Yang to explain the formation of their small CeO_2 nanocubes. A similar mechanism was proposed by Adschiri et al. [41] to explain the synthesis of CeO_2 nanocubes by a supercritical hydrothermal process. The preferential growth along the [111] direction implies a preferential deposition of monomers in this direction.

The two-phase approach promotes good control of the nanocrystal morphology by monitoring the concentration of the stabilizing agent. Figure 6.3 depicts the HRTEM characterization of CeO_2 nanocrystals processed by the two-phase approach by keeping the water/toluene ratio and the time and temperature of the synthesis constant and changing the [oleic acid]/[Ce^{+3}] ratio where the oleic acid is the stabilizing agent. At a high [oleic acid]/[Ce^{+3}] ratio, a truncated cube morphology is apparent with (110) and (100) exposed facets (Fig. 6.3a). The morphology observed by HRTEM is confirmed by image simulation (Fig. 6.3b) which facilitated the 3D reconstruction morphology illustrated in Fig. 6.3c. The synthesis with an intermediated [oleic acid]/[Ce^{+3}] ratio produced CeO_2 nanocrystals with a polyhedral shape (Fig. 6.3d). Figure 6.3e and f depict the image simulation and the 3D reconstruction image of the polyhedral nanocrystal with the (111) and (100) exposed facets. At a low [oleic acid]/[Ce^{+3}] ratio, the nanocrystal did not show a well-defined shape (see Fig. 6.3g). Oleic acid was also fundamental to promote the high colloidal stability of those nanocrystals in a non-polar solvent such as toluene (see Fig. 6.3h). This poly-functional property of oleic acid is related to the molecular structure of the molecule (see Fig. 6.3i). The carbonyl group (−COOH) of this molecule must interact with the CeO_2 surface to control the growth process and colloidal stability.

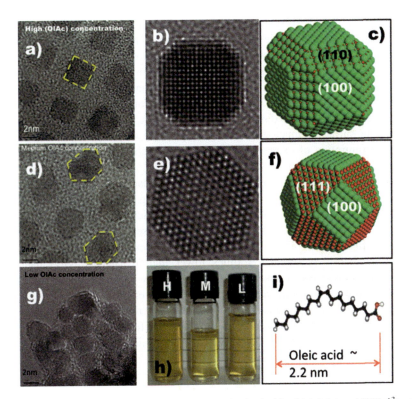

Fig. 6.3 (**a**) HRTEM image of CeO_2 nanocrystal synthesized with a high [oleic acid]/[Ce^{+3}] ratio (H); (**b**) image simulation of the truncated cube; (**c**) 3D reconstruction of the truncated cube; (**d**) HRTEM image of a CeO_2 nanocrystal synthesized with a medium [oleic acid]/[Ce^{+3}] ratio (M); (**e**) image simulation of the polyhedral morphology; (**f**) 3D reconstruction of the polyhedral shape; (**g**) HRTEM image of a CeO_2 nanocrystal synthesized with a low [oleic acid]/[Ce^{+3}] ratio (L); (**h**) photograph of vials showing the high colloidal stability of CeO_2 nanocrystals synthesized in several [oleic acid]/[Ce^{+3}] ratios; (**i**) oleic acid molecule (*red balls* = oxygen; *white balls* = hydrogen; *black balls* = carbon)

6.2 Trends in the Application of Metal Oxides with Controlled Shapes and Reactive Surfaces

In the last 5 years, metal oxide nanocrystals with tailored facets have attracted intense research interest. However, their application as commercial products is not practical yet. The main reason for that is the lack of a chemical protocol that allows reproducibility and large-scale production. However, we can predict several applications of metal oxide nanocrystals with tailored facets. Here we will describe some examples of the application of TiO_2 nanocrystals with highly reactive facets. We select TiO_2 because it is an abundant, low-cost, and environmentally benign material. Besides, TiO_2 has obtained commercial success and it is widely used as white pigment.

The pioneer synthesis of anatase TiO_2 crystals with {001} facets, reported by Lu et al. [13, 14], allowed the development of materials with superior performance for several applications. For instance, the anatase nanosheets with dominant {001} facets showed superior activity for photodegradation of small organic molecules, photocatalytic water splitting, and excellent performance in photosensitized anode materials for solar cells [42]. These superior performance properties related to surface chemistry can be directly linked to the fact that reduction and oxidation sites on the surface of anatase TiO_2 single crystal are spatially separated because of selective migration of excited electrons (negative charged) and holes (positively charged) [42].

The anatase nanosheets with dominant facets also showed interesting performance as anode materials for Li^+ ion batteries. Using anatase with largely exposed {001} facets, the Li^+ ion insertion/extraction kinetics and batteries' performance were investigated by Yang and co-authors [43]. These authors reported an irreversible capacity loss for TiO_2 nanosheets with 62% exposed {001} facets of only 10.7%, which is more than 3 times lower than of anatase TiO_2 nanotubes. Moreover, the nanosheets showed excellent capacity retention after several cycles of Li^+ intercalation/de-intercalation process. However, the nanosheets present low electronic conductivity and lattice strains induced by repetitive intercalation cycles, which are important problems to be overcome. Therefore, strategies must be used to improve the performance of this anode material in Li ion batteries.

References

1. Seyed-Razavi, A., Snook, I.K., Barnard, A.S.: Origin of nanomorphology: does a complete theory of nanoparticle evolution exist? J. Mater. Chem. **20**, 416 (2010)
2. Liu, G., Wang, L., Yang, H.G., Cheng, H.-M., Lu, G.Q.: Titania-based photocatalysts—crystal growth, doping and heterostructuring. J. Mater. Chem. **20**, 831 (2010)
3. Halder, A., Kundu, P., Viswanath, B., Ravishankar, N.: Symmetry and shape issues in nanostructuregrowth. J. Mater. Chem. **20**, 4763 (2010)
4. Xie, X., Li, Y., Liu, Z.-Q., Haruta, M., Shen, W.: Low-temperature oxidation of CO catalysed by Co_3O_4 nanorods. Nature **458**, 746 (2009)
5. Xie, X., Shen, W.: Morphology control of cobalt oxidenanocrystals for promoting their catalytic performance. Nanoscale **1**, 50 (2009)
6. Deng, W., Flytzani-Stephanopoulos, M.: On the issue of the deactivation of Au–ceria and Pt–ceria water–gas shift catalysts in practical fuel-cell applications. Angew. Chem. **118**, 2343 (2006); Angew. Chem. Int. Ed. **45**, 2285 (2006)
7. Si, R., Flytzani-Stephanopoulos, M.: Shape and crystal-plane effects of nanoscale ceria on the activity of au-CeO_2 catalysts for the water–gas shift reaction. Angew. Chem. Int. Ed. **47**, 2884 (2008)
8. Ferroni, M., Carotta, M.C., Guidi, V., Martinelli, G., Ronconi, F., Sacerdoti, M., Traversa, E.: Preparation and characterization of nanosized titania sensing film. Sens. Actators B **77**, 163 (2001)
9. Zhang, Z., Wang, C.C., Zakaria, R., Ying, J.Y.: Role of particle size in nanocrystalline TiO_2-based photocatalysts. J. Phys. Chem. B **102**, 10871 (1998)
10. Hagfeldt, A., Graetzel, M.: Molecular photovoltaics. Acc. Chem. Res. **33**, 269–277 (2000)

11. Lazzeri, M., Vittadini, A., Selloni, A.: Structure and energetics of stoichiometric TiO_2 anatase surfaces. Phys. Rev. B **63**, 155409 (2001)
12. Diebold, U.: The surface science of titanium dioxide. Surf. Sci. Rep. **48**, 53 (2003)
13. Yang, H.G., Sun, C.H., Qiao, S.Z., Zou, J., Liu, G., Smith, S.C., Cheng, H.M., Lu, G.Q.: Anatase TiO_2 single crystals with a large percentage of reactive facets. Nature **453**, 638 (2008)
14. Yang, H.G., Liu, G., Qiao, S.Z., Sun, C.H., Jin, Y.G., Smith, S.C., Zou, J., Cheng, H.M., Lu, G. Q.: Solvothermal synthesis and photoreactivity of anatase TiO_2 nanosheets with dominant {001} Facets. J. Am. Chem. Soc. **131**, 4078 (2009)
15. Han, X., Kuang, Q., Jin, M., Xie, Z., Zheng, L.: Synthesis of titania nanosheets with a high percentage of exposed (001) facets and related photocatalytic properties. J. Am. Chem. Soc. **131**, 3152 (2009)
16. Niederberger, M., Colfen, H.: Oriented attachment and mesocrystals: Non-classical crystallization mechanisms based on nanoparticle assembly. PCCP **8**, 3271 (2006)
17. Zhang, J., Huang, F., Lin, Z.: Progress of nanocrystalline growth kinetics based on oriented attachment. Nanoscale **2**, 18 (2010)
18. Polleux, J., Pinna, N., Antonietti, M., Niederberger, M.: Ligand-directed assembly of preformed titania nanocrystals into highly anisotropic nanostructures. Adv. Mater. **16**, 436 (2004)
19. Polleux, J., Pinna, N., Antonietti, M., Hess, C., Wild, H., Schlogl, R., Niederberger, M.: Ligand functionality as a versatile tool to control the assembly behavior of preformed titania nanocrystals. Chem. Eur. J. **11**, 3541 (2005)
20. Niederberger, M., Pinna, N.: Metal Oxide Nanoparticles in Organic Solvents-Synthesis, Formation, Assembly and Application. Springer, London (2009)
21. Da Silva, R.O., Gonçalves, R.H., Stroppa, D.G., Ramirez, A.J., Leite, E.R.: Synthesis of recrystallized anatase TiO_2 mesocrystals with Wulff shape assisted by oriented attachment. Nanoscale **3**, 1910 (2011)
22. Theppaleak, T., Tumcharern, G., Wichai, U., Rutnakornpituk, M.: Synthesis of water dispersible magnetite nanoparticles in the presence of hydrophilic polymers. Polym. Bull. **63**, 79 (2009)
23. Nath, S., Kaittanis, C., Ramachandran, V., Dalal, N.S., Perez, J.M.: Synthesis, magnetic characterization, and sensing applications of novel dextran-coated iron oxide nanorods. Chem. Mater. **21**, 1761 (2009)
24. Li, Z., Wei, L., Gao, M., Lei, H.: One-pot reaction to synthesize biocompatible magnetite nanoparticles. Adv. Mater. **17**, 1001 (2005)
25. Kyoungja Woo. Hong, J.: Surface modification of hydrophobic iron oxide nanoparticles for clinical applications. IEEE Trans. Magn. **41**, 4137 (2005)
26. Lu, Y., Yin, Y.D., Mayers, B.T., Xia, Y.N.: Modifying the surface properties of superparamagnetic iron oxide nanoparticles through A Sol−Gel approach. Nano Lett. **2**, 183 (2002)
27. Lin, C.A.J., Sperling, R.A., Li, J.K., Yang, T.Y., Li, P.Y., Zanella, M., Chang, W.H., Parak, W.G.J.: Design of an amphiphilic polymer for nanoparticle coating and functionalization. Small **4**, 334 (2008)
28. Talelli, M., Rijcken, C.J.F., Lammers, T., Seevinck, P.R., Storm, G., van Nostrum, C.F., Hennink, W.E.: Superparamagnetic iron oxide nanoparticles encapsulated in biodegradable thermosensitive polymeric micelles: Toward a targeted nanomedicine suitable for image-guided drug delivery. Langmuir **25**, 2060 (2009)
29. Kim, S.B., Cai, C., Sun, S., Sweigart, D.A.: Incorporation of Fe_3O_4 nanoparticles into organometallic coordination polymers by nanoparticle surface modification. Angew. Chem. Int. Ed. **48**, 2907 (2009)
30. Insin, N., Tracy, J.B., Lee, H., Zimmer, J.P., Westervelt, R.M., Bawendi, M.G.: Incorporation of iron oxide nanoparticles and quantum dots into silica microspheres. ACS Nano **2**, 197 (2008)
31. Li, X.H., Zhang, D.H., Chen, J.S.: Synthesis of amphiphilic superparamagnetic ferrite/block copolymer hollow submicrospheres. J. Am. Chem. Soc. **128**, 8382 (2006)

32. Park, J., Yu, M.K., Jeong, Y.Y., Kim, J.W., Lee, K., Phan, V.N., Jon, S.: Antibiofouling amphiphilic polymer-coated superparamagnetic iron oxide nanoparticles: synthesis, characterization, and use in cancer imaging *in vivo*. J. Mater. Chem. **19**, 6412 (2009)
33. Wan, S., Huang, J., Yan, H., Liu, K.: Size-controlled preparation of magnetite nanoparticles in the presence of graft copolymers. J. Mater. Chem. **16**, 298 (2006)
34. He, H., Zhang, Y., Gao, C., Wu, J.Y.: 'Clicked' magnetic nanohybrids with a soft polymer interlayer. Chem. Commun. **45**, 1655 (2009)
35. Shen, L.F., Laibinis, P.E., Hatton, T.A.: Bilayer surfactant stabilized magnetic fluids: Synthesis and interactions at interfaces. Langmuir **15**, 447 (1999)
36. Gonsalves, R.H., Cardoso, C.A., Leite, E.R.: Synthesis of colloidal magnetitenanocrystals using high molecular weight solvent. J. Mater. Chem. **20**, 1167 (2010)
37. Gonsalves, R.H., Schreiner, W.H., Leite, E.R.: Synthesis of TiO_2 nanocrystals with a high affinity for amine organic compounds. Langmuir **26**, 11657 (2010)
38. Skorodumova, N.V., Baudin, M., Hermansson, K.: Surface properties of CeO_2 from first principles. Phys. Rev. B **69**, 075401 (2004)
39. Yang, S., Gao, L.: Controlled synthesis and self-assembly of CeO_2 nanocubes. J. Am. Chem. Soc. **128**, 9330 (2006)
40. Dang, F., Kato, K., Imai, H., Wada, S., Haneda, H., Kuwabara, M.: Characteristics of CeO_2 nanocubes and related polyhedra prepared by using a liquid–liquid interface. Crys. Growth Des. **10**, 4537 (2010)
41. Zhang, J., Ohara, S., Umetsu, M., Naka, T., Hatakeyama, Y., Adschiri, T.: Colloidal ceria nanocrystals: A tailor-made crystal morphology in supercritical water. Adv. Mater. **19**, 203 (2007)
42. Fang, W.Q., Gong, X.-Q., Yang, H.G.: On the unusual properties of anatase TiO_2 exposed by highly reactive facets. J. Phys. Chem. Lett. **2**, 725–734 (2011)
43. Sun, C.H., Yang, X.H., Chen, J.S., Li, Z., Lou, X.W., Li, C., Smith, S.C., Lu, G.Q., Yang, H.G.: Higher charge/discharge rates of lithium-ions across engineered TiO_2 surfaces leads to enhanced battery performance. Chem. Commun. **46**, 6129–6131 (2010)

Index

A
Ab initio, 11
Acetone, 86
Activation energy, 21, 29, 33, 62
Activities, 9, 25, 37, 83, 90
α-Al_2O_3, 23
γ-Al_2O_3, 23
Anatase, 11, 21, 23, 49, 64, 76–78,
    84–86, 90

B
$BaTiO_3$, 87
Benzyl alcohol, 84
Bubble model, 20–22
Building blocks, 45, 64, 83, 85

C
$CaCO_3$, 73, 74
CdSe, 21, 58
Cerium oxide (CeO_2), 52, 74, 75, 83,
    84, 87–89
Chemical potential, 8, 9, 15, 19, 24–26
Clausius inequality, 8
Co_3O_4, 83
Coalescence, 45, 53–64, 70, 71, 73, 84
Coarsening, 19, 36, 59
Collision, 29, 45–48, 50, 52–56,
    62–65, 69
Collision cross-section, 50, 51, 57,
    62, 63
Colloid, 12, 15, 16, 38, 46, 64
Colloidal coagulation, 12
Colloidal dispersion, 7, 15, 16, 50
Colloidal process, 3
Colloidal stability, 13, 15, 16, 19, 87–89

Colloidal state, 12, 13, 46–65
Critical radius, 3, 22
Critical size, 3, 22, 27, 29, 31, 76, 77
Crystallization, 2, 4, 7, 19–39, 73, 74,
    84, 85, 87
Cumulative torque, 70, 71
Curie, 11

D
Debye force, 13
Debye-Huckel, 14
Diffusion coefficient, 30, 56
Dipole-charge, 13, 14, 50
Dipole-dipole, 13, 14, 49, 50
Dipole interactions, 14, 49, 50, 64
DLVO theory, 13
Driving force, 1, 20, 24, 26–28, 31, 32,
    45, 49, 65, 72

E
Effective collision, 45–47, 50, 52, 65
Electrostatic interaction, 13
Energy barrier, 22, 23
Enthalpy, 8, 23, 24, 76
Entropy, 7, 8, 15, 23, 24
Equilibrium shape, 11

F
Fe_3O_4, 34, 35, 85, 86
Fick's law, 30, 37, 38, 56
First law, 7, 8, 30, 37, 38, 56
Flocculated state, 12, 15, 16, 46,
    48–53
Fugacity, 9

E.R. Leite and C. Ribeiro, *Crystallization and Growth of Colloidal Nanocrystals*,
SpringerBriefs in Materials, DOI 10.1007/978-1-4614-1308-0,
© Edson Roberto Leite and Caue Ribeiro 2012

G

Gadolinium-doped cerium oxide
    (GCO), 52, 53
Gel point, 13
Gibbs free energy, 3, 7–9, 22–24,
    34, 75
Grain boundary, 45, 52, 69, 72–76
Grain-rotation-induced grain coalescence
    (GRIGC), 69–74
Growth, 2–4, 7, 19–39, 45–54,
    57–59, 61, 62, 64, 65, 69–74,
    76–78, 84, 88
Growth process, 3, 4, 7, 19, 20, 30,
    31, 34, 36, 45, 47–52, 59, 64,
    69, 71–74, 88

H

Hamaker constant, 14
Heat, 2, 7, 32, 49
Helmholtz free energy, 8
Heterogeneous nucleation, 33–34
High-energy facets, 46
Homogeneous nucleation, 22–28

I

Interactive cross-section, 50, 51
Interface, 12, 29, 30, 39, 46, 69–78
Internal energy, 7
Interparticle forces, 13
Interparticle potential, 13
Isoelectric point, 15

K

Keeson force, 13

L

Laplace-Young equation, 21, 36
London dispersion, 13

M

Maxwell-Boltzmann statistics, 29, 54
Mesocrystals, 4, 45–65, 74, 75, 83, 85
Metastable, 23, 28, 76–78
Microwave, 49, 50
Mismatch angle, 71
Molecular weight, 85, 86
Monomer, 3, 4, 29, 30, 32, 70, 84, 88

N

Nanocrystals, 1, 3, 4, 7, 11, 15–17, 19,
    21, 32, 34, 45–53, 58, 62, 64, 65,
    69, 70, 73–78, 83–89
Nanoparticles, 1, 2, 19, 22, 28, 32–36,
    38, 45–47, 50, 52–56, 58, 64, 65,
    70, 74, 75, 77, 83–90
Nanorods, 36, 45, 52, 64, 73, 74, 77
Nanosheets, 84, 90
Nanowires, 36, 49, 64, 74, 75, 84
Niobia (Nb_2O_5), 54, 73
N,N-dimethylformamide (DMF), 86
Non-classical crystallization, 4, 84
Non-effective collision, 46
Nucleation, 2–4, 19–39, 65, 69, 74,
    76, 88
Nucleation rate, 19, 28–29
Nucleus, 19, 22, 24, 25, 27, 30, 32,
    76, 84

O

Oleic acid (OAc), 88, 89
One-dimensional (1D), 46–49, 74
Oriented attachment (OA), 4, 36, 45–65,
    69–78, 84, 85
Ostwald ripening (OR), 36–39, 45,
    58, 59, 69, 77
Ostwald step rule, 23, 76

P

Peroxo complexes of titanium (PCT), 78
pH, 15, 16, 19, 25–28, 32
Poly(1,4-butanediol), 85, 86
Polyaniline, 87
Polydispersivity, 28
Polyethylene glycol, 85
Polyol, 85
Pt nanocrystals, 69

R

Recrystallization, 2, 52, 69, 74–76, 85
Repulsive forces, 46, 48, 49
Rocksalt, 21
Rutile, 21, 23, 76–78

S

Saturation activity, 25, 37
Sb-doped SnO_2, 47, 48

Sb:SnO$_2$, 32–34
Second law, 7, 8
Self-assembled, 46
Self-assembly, 3, 64
Self-organization, 45, 49, 64, 84
Self-recrystallization, 52, 69, 74–76
Si, 28, 34, 36
Single crystal, 1, 2, 10, 49, 71, 72, 74,
      75, 85, 90
Sintering, 3
Solid-solid interface, 69–78
(Sr,Ba)TiO$_3$, 87
SrTiO$_3$, 87
Standard chemical potential, 8, 25
Steric stabilization, 16
Surface energy, 3, 9–11, 21, 22, 24, 25,
      33–36, 46, 48, 59, 70–72, 74–78,
      84, 87, 88
Surface entropy, 23
Surface free energy, 7, 11, 22, 23
Surface relative energies, 84

T
Tetrahydrofuran (THF), 86
Tin oxide (SnO$_2$), 10, 15, 16, 32–34, 47,
      48, 53–55, 58, 62, 64

Titanium (IV) oxide (TiO$_2$), 11, 21, 23,
      32, 36, 47, 49, 54, 59, 64, 76, 78,
      84–87, 89, 90
Transition metal oxides, 83–89

V
van der Waals (vdW), 13, 48
van der Waals (vdW) forces, 48, 49

W
Work, 4, 7, 9, 10, 20, 22, 24, 28, 34, 36,
      47, 49, 64, 73, 76
Wulff construction, 3, 11, 48, 87
Wulff, G., 11
Wulff rule, 11
Wulff shape, 11, 85
Wurtzite, 21

Z
Zero-dimensional (0D), 46–48
Zeroth law, 7
Zeta potential, 15, 16
Zirconium oxide (ZrO), 23, 32, 65, 76–78
ZnS, 54